The Beekeepers Annual 2017

THE BEEKEEPERS ANNUAL
IS PUBLISHED BY
NORTHERN BEE BOOKS
MYTHOLMROYD,
WEST YORKSHIRE
PRINTED BY
LIGHTNING SOURCE, UK
ISBN 978-1-908904-92-8
MMXVI

SET IN HELVETICA LT BY D&P Design and Print

Cover: Autumn in a Ukrainian Apiary, painted by *Nikolay Yarovoy*, Sumi, Ukraine.

EDITOR, JOHN PHIPPS
NEOCHORI, 24024 AGIOS NIKOLAOS,
MESSINIAS, GREECE
EMAIL manifest@runbox.com

The Beekeepers Annual 2017

CONTENTS

FOREWORD

John Phipps

September 2016

At the time of publication of the 2017 Beekeepers Annual, round about the 11th of November, we will be remembering all those service men and women who lost their lives in the two World Wars. 2016 marked the centenary of some of the heaviest fighting in WW1, especially during the Battle of the Somme, on the Western Front, which resulted in the loss of thousands of soldiers' lives and innumerable casualties, throughout the late summer and early autumn months. Fortunately, most of us cannot even imagine the sheer horror, discomfort and deprivation of the soldiers in the trenches during those years of war, but surprisingly, during their time away from the front line, or as they moved from place to place to be billeted in strange villages, they found firm friendship amongst the Belgian and French people. Those who were beekeepers and had left their bees in England, and who were no doubt worried about the Isle of Wight Disease which was wiping out their apiaries left in other people's hands, were to come across colonies where even amongst the shells and bullets, they derived both pleasure and solace to be in contact with the insects that meant so much to them. Of course, too, they came across bombed-out apiaries and pillaged hives, the wanton destruction of healthy colonies made more sickening when so many bees were dying because of disease in England. For many months, Lance Corporal (later Sgent,) A G Atwell, sent lengthy reports of his experiences to the British Bee Journal whist serving with his regiment, the 1st South Wales Borderers, as a stretcher bearer and medic helper between the outbreak of the war and 1916. I have collected all the reports and together they make up "A Soldier's Tale" which gives us not only an eyewitness account of those dreadful times but also shows his keenness to learn about the local bees and to salvage their produce despite the dangers around him.

As you sit in front of your warm fire this winter, are you content that your bees are safely tucked up with plenty of stores, headed by young queens, free from disease, and are likely to be alive and well in spring? Have you gone by the book and given them all the attention that they needed during the season, or are you in the words of Ian Copinger a 'coarse beekeeper' whose dedication and attention to fine detail is much like those who indulge in 'other intricate hobbies such as piano smashing', for example? Read his National Honey Show prize-winning essay to find out for yourself.

Former farmer, using all the pesticides and herbicides at his disposal, hunter, filler-in of ponds and remover of hedgerows, Bill Clark, changed after his marriage to that of a countryman whose chief interest lay in conservation. This led eventually to the wardenship of a nature reserve in Cambridgeshire and the education of people of all ages both in the field and via the radio and printed media. He poses the question 'Can animals and birds recognise human faces?' - and if so, does this also apply to a beekeeper's bees? I certainly believe that many animals and birds learn quite quickly whether humans are a threat to them - and bees, too, if kept in the garden, get used to the beekeeper's presence and are unlikely to cause any harm.

Bees have a timeless connection with the deities of many ancient cultures and as such they are revered and seen to have mystical powers. Not surprisingly then, the bee has become an important symbol throughout history and has been used by many dynasties from ancient Egypt up to Napoleonic times, by masonic organisations, and by religious bodies including the Catholic Church. Andrew Gough, who has a great interest in this aspect of beekeeping, explores in some detail the power associated with bees in mythology and the reverence and adoration which has been paid to them throughout history.

For many beekeepers 2016 was not a good season for beekeeping. Some say that it was as bad as the very poor one of twenty years ago. Here in Greece, the season has been the worst for over forty years with hardly a drop of honey in the hives and bees struggling on just a few combs. Should there be a large population of wasps or hornets this autumn, many colonies will find it difficult to survive their attacks. Let's hope for stronger colonies and better harvests in 2017.

John Phipps
Editor
31st August 2016.

Sgent. A G Atwell

A SOLDIER'S TALE:

In 1915, a honeybee queen sent through the post to the editors of the British Bee Journal, Thomas Cowan FGS FLS FRM and William Herrod-Hempsall FES, along with the letter below, resulted in frequent reports from Lance-Corporal (but later, Sgent) A G Atwell describing his experiences on the Western Front between 1914 -1916.

You will no doubt be rather surprised to receive a queen bee so early in the year as this, and perhaps be more surprised still when you find that she has come from " the Front." As a matter of fact, I've saved her from the Germans. Her eleven other sisters on their big frames, and the hives, have been blown to pieces. The hive I have taken this one from was badly knocked about, all the tops of the frames being exposed to the weather and most of the honey granulated with the cold, yet in such adverse circumstances as these she had a nice patch of brood and eggs, with not a trace of disease so far as I could see. To have left the bees there would mean simply leaving them to die, either by exposure to the weather or the visit of other German shells, which have demolished every house in the village and also the fine old church, which has been a special target for the Huns. I know I am sending you the queen at a rather awkward time, but I thought that (if she arrives alive) you may find some way of preserving her and thus save the last remnant of the fine old French apiary. The strain of bees must be very prolific, as they have such huge brood frames to keep going. I shall have more to tell you of this apiary later.

(Unfortunately our correspondent placed a small piece of comb containing honey in the box with the bees, and during their journey they became covered with the honey. This, coupled with the cold, was too much for them. The queen was just alive and also five of the workers, but the former was dead ten minutes after the box was opened, although she was promptly warmed near the fire. The five workers became quite lively.Ed)

France, 23/8/15. BBJ 2nd September, 1915

I have had so many experiences and seen so many sights connected with our art during this awful war that I feel I should not be doing my duty towards my fellow bee-keepers at home were I to remain silent on the subject. I am no journalist, and were it not for the extraordinary nature of my adventures in this country and Belgium, I doubt if I should ever have attempted to write an article for publication. However, I apologise at once for any weakness I may show in this direction, and rely upon the novelty of my story (all of which is absolutely true) to compensate anyone who should think it worth while reading. Besides things apicultural I intend to give some of my military experiences also, as the combination will enable me to explain things more fully, besides making a much more interesting story. Where possible I have given the dates and names of places, although I have not always been able to do this, owing, sometimes, to not having made the proper entries in my diary, and at other times because it would be forbidden for military reasons. If I had been in possession of a camera I could have taken some most interesting photographs, but as such a thing is not allowed by the authorities, I have done the next best thing and brought away many souvenirs. Some of these consisted of pieces of heavy shells, etc, and it was with great difficulty that I managed to carry them about with me for several months before I could get them safely home. I had originally intended to relate my experiences after the war, but as the end seems still a long way off, and as I feel that my story would prove more attractive if written in the trenches, I have decided to do it "right now," as the Americans say, and only trust that my adventures will be as interesting on paper to my readers as they were in reality to myself.

It may be as well to say that I am writing these lines within a hundred yards of the German positions, where we are using an old cellar (all that remains of a house) for our first-aid regimental dressing station, of which I am in charge. The troops in one of our front line trenches are only a few yards from the Germans, and can shout to one another quite easily. Continuous fighting with bombs, hand grenades, and trench mortars is going on there, and I am continually being called away, even while writing these lines, to attend someone who has been hit, often finding them past any aid that can be rendered them in this world. This is a spot which has seen some of the most severe fighting, and was actually in the hands of the Germans a few months ago.

It is just over twelve months ago now that war was declared by England against Germany, and yet the suspense of those few awful days which preceded the declaration hangs in my memory as if it were yesterday and ranks in my mind with some of the worst which I have experienced out here. Of course, domestic affairs naturally came first, but next to them the bees, and as it is the bees that I intend writing about, I will leave other matters out. When the war came I had sixteen stocks all nicely fixed up and bringing in loads of honey, but being a reservist of the first class I was called up immediately, and had to leave all my bees just as I was beginning to reap the benefit of my work of the previous spring and winter. However, I resolved that when I got to France I would keep my eyes open and see what bee-keeping was like on that side of the Channel. That my observations have not been in vain I think the following story will prove, and although most of my adventures have been attended by some risk, it has really made them more exciting, and has often passed away a few weary hours and turned my mind, for a little while at least, from the horrors of this terrible war.

It was on August 12th, 1914, that I embarked with my regiment, the 1st South Wales Borderers. We disembarked without mishap, and after a rest of a few hours we marched through the town and arrived at our camp on the top of the cliffs at 5.30 the following morning. Although it was only just breaking day as we came through the town, the windows above the shops and cafés were crowded with people , most of them still clad in their night attire, but anxious to catch their first glimpse of the English soldiers and to shout out a welcome, which at the time very few of us could understand. Of the few days that we stayed at the camp I have very little to say, only that the weather was extremely hot and we experienced a most violent thunderstorm one night which threatened to carry our tents away. The people also made a great fuss of us and treated us well during our few days' stay there.

We entrained again on Saturday, 15th August and made a twenty hours' journey in cattle trucks. This was naturally very trying, but the novelty of the country which we passed through and the fact that we were going to meet the Germans made us forget, to a certain extent, the rather trying experience of representing a box of sardines. We were not sorry the following day (Sunday) to arrive at our destination, a little village in northern France called Etreause. Here, after disentangling ourselves, we all indulged in a good stretch which enabled our bodies to once more assume their natural proportions, and then after a rest of a few hours, during which time many enjoyed a bathe in a swift little trout stream, we marched eight or nine miles in the most drenching rain that I have ever experienced to another little village called Leschelle. It was during our few days' stay at this place that I caught sight of the FIRST FRENCH APIARY, if it could be called such. It was really a two-storied bee house, such as I afterwards found to be quite common in this part of the country, and which contained several skeps of bees, which were working well.

BBJ October 7th, 1915

I also saw a long row of straw-covered skeps in the same village later. But unfortunately I had no time for further investigation, as we were kept rather busy with our preparations to meet the Germans, and could I have met the owners of the bees I am afraid I should have found great difficulty in making myself understood, as my knowledge of the French language was not at its best just then. I may mention that it was on the night of the 19th August that the first post arrived from home, and there were very few of us who did not get a letter.

Leaving Leschelle on the morning of the 20th we marched the greater part of the day, reaching a little village called Malgarni, but our stay here was very short, for we were off again early next morning, and were now well on our way to meeting the Germans, who, of course, as everyone knows, were rapidly invading French and Belgian territory. After another long march in very hot weather we arrived at the village of St. Aubm, another short rest and we were off again early next morning; we had no time to lose then. Often as we plodded along the dusty roads we could see the tops of a row of rustic skeps sticking above the hedges, and a village would never be passed without I could see one or more of these old-fashioned little apiaries clustered under the hedge or standing under the shelter of the overhanging thatched roof of one of these quaint old whitewashed cottages.

I must not forget to mention the thousands of refugees we were continually passing on the road. These were of all ages, from babies in arms to old men and women of some 80 or 90 years of age, some plodding along carrying a few of their belongings as best they could, others more fortunate were driving along with all the furniture and accoutrements that it was possible for them to pack on their much overloaded wagons; these vehicles ranged from the little cart drawn by a dog to huge wagons to which were harnessed a team of four or six oxen. Everything that had wheels seemed to be in evidence. On many occasions I have seen these old wagons (many of which had not been used for years) break down with their great loads, and the people have had to leave them where they lay, for there was no time to lose.

The booming of the bursting German shells could be heard in the distance, and the black clouds of smoke which could just be seen rising over the horizon meant that another village was being sacrificed to the flames, striking terror to the hearts of the refugees, and spurring the soldiers on to meet this inhuman foe. I think the sight of these miles of fleeing people was one of the saddest that it has ever been, or ever will be, my lot to witness, and I feel it quite beyond me to adequately describe the scene on paper. One little thing more, however. The people were driving almost every living thing from their farms before them—cows, sheep, and pigs; chickens and ducks packed in baskets and slung under the wagons; all were there. But the bees? They could not very well bring the bees! And so these little apiaries with their healthy workers and their year's stores still intact were left to the mercy of the invaders.

Well I remember about this time passing a house, the owners of which were just preparing to leave. Their bundles were ready on the doorstep, and they were only waiting for the master of the house, who was in the garden having a last look at his bees. He had some six to eight large frame-hives there, which, by the way, were supered and must have contained a considerable amount of honey. This was the biggest apiary of frame-hives that I had yet seen, and by their appearance the old man must have given a lot of time to them. I should like to have spoken to him, but as we were only marching by I had not the time, but I often wondered afterwards if he ever left his bees.

BBJ October 28th, 1915

In the afternoon we reached Mauberge here we rested at the roadside for about four hours. It was at this village that we received our first pay, although we needed very little money, as the people would give us anything. We moved on again to a somewhat larger village for the night, the name of which I have not entered in my diary. We slept in the village square, most of us. The place was also crowded with refugees, who were sleeping for the night anywhere they could find room to lie down. I remember a few wounded being brought to the village: they were chiefly cavalry scouts, and had been wounded in a skirmish with the German outposts.

We moved a little further on the next day, and got into action for the first time. We held out as long as possible against the advancing hordes of Germans, but on Monday, the 24th of August, we were compelled, in the face of greatly superior numbers, to start our retirement. Of course, this was at about the same time that the retirement from Mons took place, our brigade being on one of the flanks and not actually at Mons itself. We were fighting rearguard actions every step, and had some very heavy marching. We passed through the town of Soissons on Monday, the 31st. We reached the farthest point of our retirement at about 5 o'clock on September 2nd, having been pressed back to within a few miles of Paris. In fact, rumours were current among the troops that we were actually going there to help hold the forts. However, I am glad to say this was not necessary; for, as everyone knows, we turned about and drove the Germans right back to their positions on the Aisne.

During this advance to the Aisne I made the following notes: Started our advance on Thursday, September 3rd; that evening, while resting, was shaved by a lady barber, a refugee. For several days after this I made no entries, but if I remember rightly it was nothing but continual marching, with scarcely any time for food or sleep.

On the 8th we passed a lot of our cavalry, men and horses who had been killed while fighting the German rearguard. This commenced the battle of the Marne. For several days we still continued the advance. All the villages we passed through now had been looted by the Germans; shops broken open, wine cellars looted, cattle and poultry killed and eaten, their entrails being thrown about the ground.

It was on the 12th of September that I came across the first looted apiary. Of course, I expect I had passed several such before, but being hidden behind the hedges, etc., it was not always easy to see them in passing. However, the one I mention was standing right in the open, about six yards from the road, with no hedge or fence of any description to shut it off from passing traffic. Of the nine or ten hives (or rather straw-thatched skeps) only about two seemed to be undamaged. Others were burst open, exposing both comb and honey, while others, judging by the empty stands, had been carried completely away.

The troops passed this spot pretty quickly, for, of course, the bees were not in the best of tempers. The following day I found a looted skep by the roadside, with a few bees still hovering round. Honey and comb was all gone, and the skep was most likely one of those taken from the apiary I had seen the day before; it was of wicker work, covered with mud and cow dung, and I made a rough sketch of it at the time, a drawing from which I enclose herewith.

Sketch of Skep Looted by Germans, Aug. 13, 1914.

Pictures of skep and author with beekeeper

BBJ November 25th, 1915

Our advance still continued, and many and various were the articles left on the roadside by the retreating Germans. Food tins of all descriptions, old worn out boots and clothing, helmets and broken equipment. Occasionally a few Prisoners would fall into our hands; these were chiefly stragglers who could not keep up with the main force. We also passed many large wagons and motors, some which had to be abandoned through slight defects, and others which had overturned in their haste to get away, as the roads in places were none too wide, and the ditches on either side often very deep.

One rather sad sight was the number of horses which we passed, either dead or left in a dying condition, caused by the continual work over heavy roads. It was about this time, while resting in a field, that I saw one of the best hornets' nests that I had ever seen, it was built out from the branch of a tree, and I can assure you that none of us felt inclined to meddle with these huge insects, who seemed to be ready to dart at anyone who should venture too near them. The weather about this time was rather fine for the time of year, and during the warmest part of the day I noticed numbers of bees swarming round the empty jam tins which were thrown away by the troops. These were evidently bees of looted apiaries anxious to get food of any description to make up for their lost stores. It may be interesting to note that they favoured the orange marmalade, the tins from this conserve being absolutely black with them.

On Sunday night, September 13th, we rested in rather a big village, the name of which I think was Boulne. All along the advance the troops had been anxious to clash with the enemy, but none of us thought that the following morning would see us launched in a most terrible fight, and that many of our comrades who laughed and chatted round the camp fire this Sunday night had but a few hours left to them in this life. However, such was the case, for the following morning, after leaving the village a few miles behind us, we dashed right into the enemy near the tiny rustic village of Chevoy Boulne. The reader might easily imagine a fine autumn day among a beautiful hilly country, shrouded in its autumn splendour of gold and bronze in all their varying degrees. He might also imagine a little old-world village, with thatched and whitewashed cottages and farms and quaint old folk, many of whom had never been more than a dozen miles from their native village. But try as he might, unless he has actually experienced it, he will never be able to imagine the awful experience of going through such a fight. The roar of the guns, the continual whizz of the bullets, the shrapnel bursting all around; that peaceful country of yesterday was today a veritable hell. It seems marvellous that any living thing could exist at all. Of the horrible sights of the battlefield I will say but little, it could do the reader no good to hear of them, and although they can never be forgotten, they will best be left untold.

How the writer ever came through such an ordeal the One above only knows, but on this and four other occasions (of which I shall write later) I have had some of the nearest shaves that it is possible for anyone to undergo. Of course at this time there were no trenches, and dugouts as there are to-day, only such as could be hastily scratched out during the night.

BBJ December 9th, 1915

We found ourselves, after a day of continuous fighting, in the little village of Chivvy Boulne, Here we stayed for a week, my regiment helping to hold a ridge of hills just outside the village. The place was under a hurricane of shell fine both day and night, and I was kept so busy attending to the continual stream of wounded (I was employed in the first-aid dressing station at that time) that I had no time while in this place to go exploring for beehives, although I have not the slightest doubt that there were plenty about.

There is one little story which I might tell of this place, which I am sure will touch the hearts of all bee-keepers. Just opposite our dressing station lived an old lady of some seventy years or more; her husband was dead, her three sons were at the war, her married daughter lived miles away, and so this old lady—too old to leave her native village—was left to her fate. Often when I had a minute to spare during the day I would cross over, and she was always delighted to see me; she would open her little Bible, which she always carried with her, then amidst the thunder of the guns and bursting shells (any one of which might demolish the house at any moment), the old lady would turn to a picture of the Virgin Mary, and pointing above would exclaim, "La Saint Monsieur, La Saint." We had to evacuate the place after a week there, and withdraw to another village called Vendresse. But the old lady, together with a few other old people, stayed behind. I do not know what happened when the Germans came. Vendresse, a little larger and more modern than the last place, gave me many interesting experiences connected with bees. We had been here no longer than a day before one of the stretcher bearers (who had been exploring some of the vacated houses and gardens) came rushing up to me, and vigorously rubbing his nose, which was already showing signs of an unseen aerial attack, shouted out in great excitement: "Bees! Bees!" Two or three minutes' walk brought us to the garden, which contained a little apiary of some half-dozen skeps, of the same size and shape as the one I illustrated a few weeks ago. One skep was lying on the ground with a cloud of angry bees flying around. This one, my friend told me, he was carrying away when the bees gently reminded him that it wasn't quite such an easy task as he expected, and forced him to drop them, and beat as hasty a retreat as any he had yet experienced during the war.

BBJ January 6th, 1916

While we were waiting for the bees to get settled and quiet I occupied the time in making an impromptu smoker, which consisted of an empty jam tin stuffed with some old rags. It needed very little smoke to make the bees quiet, as there was plenty of honey exposed which had run from the broken combs. I found that the bees were really of a very quiet nature when properly handled, and it was not long before I had them driven from the skep. The bees themselves were practically the same as our British brown ones, and were typical of the numerous other skeps of bees which I have come across in different parts of the country, although I have at times noticed a slight variation in the colour, due no doubt to the introduction of some foreign blood.

The most noticeable feature, however, was their absolute freedom from disease. I have not found during my whole experience here a single case of disease of either foul brood or "Isle of Wight" although I have often found combs, absolutely chocolate in colour, which must have been in use for several years. The sanitary conditions of the country are extremely bad; putrid ponds and open cesspools are very numerous, and during the hot weather the bees undoubtedly visit them at times for their water supply. I have noticed, however, that the French beekeepers rarely forget to keep a supply of water in shallow pans near the hives.

Returning once more to the skep which I had been driving, I found that we had quite a nice lot of honey, which we carried back to our billet and divided, amongst the troops. The honey was of medium colour, and had a peculiar although not unpleasant taste and aroma, such as I have never found in English honey. The troops made very short work of it, as it came at a time when our rations were not quite so liberal and varied as they are to-day. Our jam issues then were very small, and there was no tinned butter issued as there is now. The doctor we had with us knew a good thing when he saw it, and used to take some of the honey with all his meals. The troops having once got a taste of the honey, like - Oliver Twist - were asking for more, and I had no rest until one by one the remaining skeps of that little apiary went the same way as the first. But still I think it was all for the best, because if left alone they would either have been destroyed by shell fire or knocked over and left to die by some of the troops who would have been afraid to actually try to take the honey from them.

The apiary was situated at the bottom of the garden, with only a wall and a pathway between it and the village church, which then had several shells through it, and by now I expect it is quite demolished, and the site of the little apiary buried in the debris.

Sketch of apiary at Vendresse

I have often read in the British Bee Journal of many novel experiences whilst driving bees, but to drive them while under shell fire is something quite new; nevertheless this is actually true, as shells and bullets were whizzing about all the day long. We did not mind the rifle bullets so much, as, being in a bit of a hollow, these little missiles usually passed too high to do much damage. But the shells dropped anywhere, especially round the church, and many times while in the act of driving I have had to run for shelter when the shelling got a bit too hot. It eventually became so risky to stay anywhere near the church for long that I carried the last two skeps to a cellar and operated on them there, although this place was not shell proof to the "coal boxes" or "Jack Johnsons" as we called them, and which were used very liberally by the Germans during the early stages of the war, It was only a little distance away that one did enter a cave killing a great number of officers and men, practically the whole headquarters staff of a certain regiment.

I bottled off two jars of honey and also procured some excellent wax, which I carried about with me for several weeks, intending to try and get them home as souvenirs, but as there was no chance to post parcels home then, and leave at that time was unthought of, I was forced to give the honey to the troops and to throw the wax away.

In the vicarage at this village I also came across a huge piece of wax which must have weighed (before a large piece was broken off) some 12 lb. or more. I also found little things lying about which told me that the occupant at one time must have kept a considerable number of bees.

Vendresse also holds some very grim recollections, for during our few weeks' stay there we were continually attacked by the Germans, and I lost many old friends and comrades, both officers and men, one sergeant in particular, who lived very near me before the war, and who was always very interested in my bees, and had intended starting himself if the war had not broken out.

A moonlight night just outside this village after a big attack I shall always remember; we went out searching for wounded, and the strange positions that some of the dead occupied made the scene most weird, and I spoke to a chap who was sitting up holding his rifle, only to find on closer examination that he had been dead some hours, evidently killed instantly in the position he then occupied.

I had a very narrow escape one Sunday evening about 5 o'clock. I had to go round to a big chateau to fetch the medical officer, who was attending a small church service in the cellars there. I was just preparing to leave the place when a huge "Jack Johnson" (weighing some half a ton) came crashing through the stables adjoining the house and landed without bursting on the lawn only a few yards from where I was standing. Had it exploded there would not have been much of me, or the chateau either, left. As it happened no one was hurt at all, although the shell passed clean through both walls of the stable just over the heads of men and horses.

BBJ February 17th, 1916

We stayed at the Convent of St. Barbara for about eight days, during which time the beautiful village church and most of the houses were destroyed by shell fire. Although the convent was very much exposed, the Germans, for some reason, did not shell it until the night we left. That awful night I shall never forget. The regiment were being relieved by the French, and by seven o'clock were well on their way back, together with the majority of our stretcher-bearers. The doctor, myself, and two or three men remained behind to look after some wounded for whom we were expecting an ambulance. However, this never reached us, for about nine o'clock the Germans started an attack on the positions taken over by the French. During this attack they seemed to remember the convent, for they rained shell after shell upon us. These were shells of the heavy type, and any direct hit would have blown up the convent and set it on fire. We simply sat there with the wounded, expecting every moment to be the last. It was an awful experience; these huge shells, which were being rained on us at the rate of two every minute, could be heard, and I might even say felt, coming from a great distance away, and it was not until after the terrific explosion (which shattered all the windows and shook the convent like a boat) that we knew that another had missed its mark.

The suspense eventually became so great that we decided that we must get away somehow, and a shell crashing through a part of the roof hastened this decision. We carried the wounded from the convent, down an open, fire-swept piece of road, and through the still burning houses of the village. We eventually reached safety, where we handed our wounded over to the ambulance, but it was well after daybreak before we reached our regiment. This experience so affected the doctor that the following day he had to be sent away to the base, and, as far as I am aware, has never been in a fit condition to return to the front.

After two days' rest we marched to Ypres, where we took part in the desperate fighting which took place towards the latter part of the year. I had the pleasure of seeing this beautiful old town with its famous Cathedral and Cloth Hall before it was much damaged by the Germans, and I also had a splendid view of the famous Cathedral of Rheims when passing through that place earlier in the war, a fact which I forgot to mention before. During the time we were at Ypres I had a very narrow escape. I was struck by a bullet in the back of the head, which then splashed against a wall near by and was picked up at the time and given me by a friend who was walking just behind. Although I carried a souvenir as big as an egg on my head for several days, the swelling eventually went away all right, and I am now none the worse, except for a little scar which will always remain with me and remind me of that lucky escape.

BBJ March 2nd, 1916

It was at the village of Zillebeke, just outside the town of Ypres, that I found the bee house to which I referred in my letter dated December 1st, 1914, and which appeared in the British Bee Journal dated 10th of the same month. At that time none of us thought the war was going to last so long, and the hope I expressed that the owner would soon be able to return to his bees I am now afraid will never be realised, for I expect by now that this little apiary has quite vanished. In the same letter I wrote about my own bees, which are now all unfortunately dead through "Isle of Wight" disease. As this case may prove interesting, I will quote a few lines from the letter. I wintered six stocks in 1913. They all came through all right, but four lots showed signs of foul brood. One lot I thought best to destroy, and another I re-hived on clean foundation. About this time—April—I noticed a number of dead bees outside the hives and a few crawlers. I sent a packet of dead bees to the British Bee Journal office, and the verdict was "Isle of Wight" disease. However, I was determined to give the bees a chance, and so I put them all in clean disinfected hives and fed them up with sugar, using nothing in the way of medicine but Apicure, which I put in all the hives. There then came a marvellous change for the better; both complaints seemed to vanish, and I took during the season 500 lbs. of extracted and 205 sections, increasing the bees from five to fourteen vigorous stocks, which were all packed down on ten frames with a huge quantity of their own stores. The ten frames were left because I was then away at the Front. They all but one stock came through the following winter all right, and when I was home for a few days' leave in February, 1915, they seemed to be in quite a normal condition. The majority of them started the spring well, and most of them were supered up and gave some honey. But both complaints now seemed to come back again in a much stronger form, and by the end of the summer (despite the disinfecting with Izal my wife gave them) they were all dead.

The disease seemed to vary with the weather, and some weeks my wife would write to say that she thought they were pulling round, and she had been able to super two or three more stocks. But a week or two later she would say how much worse they were, some of them only covering a few frames. Of course, my wife did all that was in her power to pull them round, but she was not able to put them all in clean hives, or perform other operations which I should have been able to do, and so, of course, the honey take was very small. A number of the hives still remain closed up in the orchard, and when I go home this month (having completed my military service) I hope to make a short report on them. I also intend starting again with two hives, and perhaps some bee-keeper whom the new Act affects may be in a position to supply me with them.

Returning once more to my story. We find ourselves marching back for a rest and to reorganise after our great losses, which, after all, were nothing like so bad as those of the Germans, as, of course, they did all the attacking, and were mown down by our fire as they used to advance in such hordes in their endeavour to break through to Calais. However, they never got through the thin line which at that time barred the way, and so I do not think they can have much hope of doing so now. It was at the end of November that we started our march back, passing through Dickebusch and Locre, then across the Belgian frontier into France. We passed through the town of Bailleul, and on to a little village called Outtersteine. Here we billeted at the various houses and farms, and I was fortunate enough to stay at a farm the owner of which was a bee-keeper. His farm had been damaged by a shell; all his wine, eggs, and stores had been looted by the Germans as they were retiring back through the village, being pursued by the British. However, this old bee-keeper never left his bees, and I hope to tell all about him in the next instalment of my story.

BBJ April 20th, 1916

Monsieur Emile Crinquette was the name of the old French bee-keeper whom I promised to talk about in my last instalment. I was billeted at his farm for about six weeks, as it took us a considerable time to re-organise after the hard fighting we had just been through. I regret that I was unable to obtain any photos of the old farmer, his bees, or his partly wrecked house, as they would have proved most interesting. However, we were not allowed cameras, and there were no local photographers, and so I shall have to depict the scene, as well as I can, in words alone.

Mons. Crinquette was a big, burly man of some sixty years; he lived with his wife and little niece in the quaint old village of Guttersteine, Northern France, not very many miles from the Belgian frontier. He had a good-sized farm, keeping a servant girl and three or four farm hands. It was a farm of the usual type, a dozen cows, pigs, chickens, etc., with several good-sized fields adjoining, some under the plough and some used for grazing and fruit growing. Quite close to the farmhouse, in the kitchen garden which adjoined it, were the old man's bees. A nice little spot, the apiary being sheltered by a good thick hedge; in front and around it grew various herbs, thyme, sage, etc., and, although it was in December when I saw it, I could guess what an ideal spot it was for the bees, and how pretty it would look in the summer time.

Furthermore, the big, old-fashioned kitchen windows overlooked the apiary, so that it could always be watched at swarming time, and thus very few swarms were lost. The bees were chiefly in skeps, and were sheltered under a long bee-house, the floor of which was not much higher than the ground, the roof being tiled, and the three sides thatched with straw. Some half dozen or more skeps were always wintered here, meaning, of course, a goodly number in the summer, when the bee-house would become over-crowded, and the old beekeeper would gradually extend the row of skeps from it in a straight line, under the shelter of the hedge, right down the garden. I think the old man was only a skeppist because, with so much other work to do, he had not sufficient time to give to frame hives, as he had one frame hive of the outdoor observation type, and he told me how he used to study his bees in this, through the window at the back of the hive, when he had a few moments to spare.

Things used to go on quietly year after year in the same way, until the war broke out. Then the Germans came; they must have been chiefly cavalry patrols, as I think the infantry hardly reached so far as this village, not in great numbers, at any rate. However, whatever they were, they soon made short work of all the old man's stores. Everything in the way of wine, or beer, was quickly taken, after which they helped themselves to such things as eggs, butter, honey, etc., all these things being stored in large quantities by the farmers, to tide over the winter. Still, the old bee-keeper and his wife refused to leave their farm, but they had to sit and see all their goods taken, and, I am glad to say, the behaviour of the looters towards them was better than it usually is, although, of course, they daren't say a word on behalf of their property.

The worst was yet to come, for a day or two after, a German shrapnel shell caught the farm, blowing off the kitchen roof, smashing all the windows; many of the bullets and pieces of shell passed through the kitchen, embedding themselves deeply in the opposite walls. A few spare pieces also caught the apiary, the frame hive in particular, but not sufficiently to harm the bees. Fortunately there was no one in the kitchen at the time, or they must certainly have been injured. Mons. Crinquette's coat, which was hanging on one of the doors, was absolutely riddled with bullet holes, and, as it is quite impossible for him to wear it again, he is keeping it as a souvenir—he was very thankful he wasn't in it when the shell came.

I had several talks to this old bee-keeper during the time I stayed at his farm, also to several others in the same village, the majority of whom kept their bees in skeps and bee-houses. Being the winter time, of course it was not possible to examine the bees in any way but, as far as I could gather, there was very little disease about. The old man was sorry when we had to go away, and, after embracing me, kissed me on both cheeks at parting.

BBJ May 25th, 1916

After our stay at Guttersteine had lasted some five or six weeks, we received a very sudden order to move to a new part of the line. We started our march early in the evening, and reached the town of Merville about midnight, this place being about midway between Guttersteine and our destination. We rested there for several hours until, daybreak, and then continued our march, which brought us in the afternoon to Bethune, quite a fair-sized town some ten miles from the firing line. Marching straight through here we made our way to the front, and launched an attack against the Germans at a small village called Festubert, at which place I had some very interesting experience in connection with the bees. There was also a little hamlet called Gorre, which we passed through just before reaching Festubert, at which I found a quaint old cottage and bee garden, having some eight or ten skeps wintered down, and there were quite a lot of appliances and empty skeps in a barn, which proved that the people who once lived there had done considerable business with their bees.

We spent our Christmas at Festubert, and during our stay there, when things were a bit quiet, I would often take a walk to see the bees at the cottage. Unfortunately, at each visit I would find one or two skeps fewer, sometimes caused by a shell dropping near, but in the majority of cases it was the Tommies (who were billeted in the cottage) trying to get at the honey. At the last visit I made I found every skep gone.

I was just beginning to think that my adventures with the bees around this part were finished, when one day a friend came running into the dressing station carrying two large frames of honey, which he told me he had taken from a hive at the other end of the village near the firing line. I learned from him that there were quite a lot of hives there, although many had been badly knocked about by shell fire. Having nothing particular to do at the time I got him to take me up to the garden, although we ran some risk from shell and rifle fire. The house, which stood on one side of the main street, was quite near the old church, which was a total wreck. We passed through a small front garden and then through the house (which was badly knocked about, very little of the roof remaining) into the long garden, at the back of which stood the apiary of large frame hives running right down the side of the garden. But what a sight; I could do nothing for some moments but gaze in silence on the scene before me. Only about two of the dozen or more hives seemed to have escaped destruction from the murderous fire which only a few days before had been rained upon them. Huge holes made by shells of the heaviest calibre dotted the garden. Hives were torn to pieces, broken combs and frames scattered everywhere, and little clusters of bees, the majority of which were dead through exposure to the cold, made up a scene which I feel quite unable to properly describe.

Had the bees been suffering from any disease, such as "Isle of Wight" or foul brood, matters would not have been so bad, but a thorough examination which I gave them failed to disclose the slightest trace of anything wrong. They had plenty of stores, but I had some difficulty in finding combs containing honey which were not damaged much, but I managed to find a few which I tied together to take away.

Passing back through the house I found some postcards scattered about, and by the addresses on them I came to the conclusion that the owner of the apiary was Mons. Omer Francois. I also found some photos of the family, and have every reason to believe that one of the gentlemen must have been Mons. Omer Francois. I hope to tell later of a second visit which I paid to this apiary, also of one or two other interesting experiences which I had at this village.

The illustration given is of a number of some of the souvenirs collected by Sergt. Atwell. 1. - the time fuse from a heavy shrapnel shell. 2. & 5. are shrapnel bullets; 3. a rifle bullet which wounded Sergt. Atwell in the head. It will be noticed the bullet has been knocked out of shape; 4. two pieces of moulded beeswax; 6. a shell splinter; and 7. a piece of a "Jack Johnson".

BBJ Aug 31st 1916

The next time I paid a visit to the Apiary at Festubert I found that it was practically demolished: not only had many more shells fallen around, but most of the hives which were any good for the purpose were used to help make barricades or build up trench supports, as a trench now cuts right through the apiary. There was only one stock left, and there was no roof on the hive. The honey in the outside combs had granulated with the cold, but the bees were clustered in the centre, and upon examination I found several patches of brood in the middle. This, if I remember rightly, was about the middle of March, and as I felt that the bees could exist very little longer where they were, I found the queen and sent her in a box to the BRITISH BEE JOURNAL Office, but, unfortunately she did not live long after arrival. There was so much wax lying about from the many broken combs that I decided to collect it, and melt it down. I therefore went a few days later and gathered it all up. I took it into an empty house a little distance away, where I found a big, round galvanised tank: this I half filled with water, lit a big fire under it, put all the combs into it, and boiled them well down, and then strained the contents through several thicknesses of very fine wire netting. When it was cold I took a good-sized cake of wax from the top of the water, which I remelted into a more convenient size for carrying. I now have the cake of wax among my other curios, although, of course, it really needs refining again, as it is rather dark in colour, owing to its having been procured mostly from old brood combs.

The next place I should like to talk about is the Orphanage at the town of Bethune, some eight miles behind the firing line, but of course well within range of the Huns' big guns, and they frequently bombard the cross roads, railway station, and other important parts. Once when we were resting at Bethune we were billeted at the Orphanage, as all the inmates had been sent to a safer part of the country. Only two or three of the staff remained to look after the small farm which included several stocks of bees. I soon got into conversation with them as I could see the bees wanted attention. They were very pleased when I told them I understood the bees, as the man that usually looked after them was away at the war. Some section and shallow frame racks had been left on them from the previous year, but they were badly fitted and wanted re-adjusting. This I soon accomplished, having to cut away several nice slabs of honey, which the bees were beginning to place in rather awkward positions as the racks were not properly fitted. In addition to the frame hives there was a bee house with several skeps full of bees, but I was unable to do very much to these as it was too early in the year. I paid several visits later to this apiary, and was always made welcome by the people there.

BBJ November 30th 1916

Each time that we took our turn in the trenches, I usually managed to find a few moments to spare to visit some of the apiaries which I had previously found. But what a difference there was at each visit. The few remaining hives and skeps which had marvellously escaped the hurricane of shell-fire got fewer and fewer in number until there was scarcely one live bee to be seen in any of these deserted gardens.

However, a few weeks later, as the spring advanced, I was pleased to see quite a lot of our little friends busy on the trees and shrubs which had somewhat escaped the fire of the Huns. Chief among these was the peach, with its lovely pink blossoms, with the gooseberry and currant following closely upon it. These bees, of course, had come from a safer part, back behind the firing line, where the shells had not reached them. It seemed strange to see these little insects busy gathering their stores with our own guns booming all round, and German shells bursting quite near, and which at any moment might burst in the very garden where they were so busy. They seemed, however, to hum a note of defiance as they went steadily on with their work.

Many combs were scattered about the gardens from the wrecked hives, but they did not attract many bees. It seemed as though they understood and respected this mass of wreckage—once a flourishing apiary. As I walked back through the skeleton of that which was once a house, I could not help thinking how lucky we bee-keepers really are here in England. Many of us have lost our stocks from "Isle of Wight" disease (I myself have lost the whole of mine), but we still have our hives and appliances for a fresh start, and, above all, we still have a roof over our heads. In some of my early articles I told how I had endeavoured to fix up frame-hives and skeps which I had found knocked about, and how I hoped that the owners would soon be able to return to them. This was some time ago, but the war still goes on, and I cannot see how any apiaries within the fighting zone can exist at all now.

Nothing can remain to mark the places where they once stood but a few scattered pieces of wood, once hives. However, the experiences I had among the bees at the Front during the early days of the war I shall never forget, and

I shall always treasure the many little relics that I was able to bring home with me, some of which I collected at considerable risk.

As I have now reached the end of my story, I can only hope that my efforts to describe my experiences have proved interesting to readers of the Journal. I hope to restart bee-keeping myself again next spring, and I have the consolation of knowing that whatever my misfortunes may be in the future, they can never touch those of the bee-keepers and the bees at the Front.

Soldiers slang for German incoming shells:

Jack Johnson: The largest shell used by the Germans, between sixteen and seventeen inches, which explodes making a shell crater about twenty feet deep. This shell is also called the 'Ypres Express,' as it reminds one of an express train as it tears through the air emitting a dense cloud of black smoke when it explodes.

Coal Box: A German high explosive shell similar to the 'Jack Johnson' which on bursting makes a terrific noise and eliminates a heavy black cloud of gas. Should it, however, burst too near you, you don't see the cloudy effects."

Bertha: The sixty-ton German gun, so called from Bertha Krupp, of the manufacturing firm. This gun has a range of ten to twelve miles, and throws a twelve hundred pound shell which the British soldiers also call 'Jack Johnson.

THE ART OF COARSE BEEKEEPING

Ian Copinger

National Honey Show essay 2009 - 1st prize

You must consider carefully before following the path of coarse beekeeping. Its disciples must have the same dedication and attention to fine detail as those who take up any other intricate hobby such as piano smashing.

The first steps of the coarse beekeeper are easy. Your local library will provide you with a copy of one of the many books written by an experienced beekeeper which will illustrate the equipment needed and describe in detail the life style of the honey bee. Many experienced beekeepers feel it is incumbent on them to write such a book. Do remember to renew your possession of the book at the library before fines are imposed, that would never do.

The same library may be able to put you in touch with a local beekeeping association and give you details of their meetings. You should go to a meeting and introduce yourself as being keen to learn about the craft. At this stage a demonstration of enthusiasm works wonders. It might also get you a copy of a beekeeping equipment dealers catalogue. This will save you having to contact one since none, so far as I know, have 0800 telephone numbers. Although allowing yourself to enquire generally about membership and the possibility of free beekeeping classes your enthusiasm should not allow you to actually pay a subscription.

Reading the catalogue together with the beginner's book will immediately convince you that your first pound of honey could be very expensive indeed. However, the coarse beekeeper knows that no corner must be left uncut in the search for true perfection.

Your occasional attendance at a meeting, or the hoped for classes, will allow you time to gather up the minimum amount of such essential equipment that can't be substituted by other items. A longish screwdriver and a paint scraper from your toolbox would replace a hive tool. A suitable length of net curtain worn over a broad brimmed hat and tucked well into a jacket could well replace safety equipment such as a veil. A more sophisticated version I have seen is an old fencing mask with further material sewn around it to prevent access by bees. A replacement for a smoker is more difficult unless of course you are a smoker yourself in which case a pipe, filled with well-rubbed War Horse, or a small cigar will suit admirably and, yes, I have seen it done.

At association meetings always listen for mention of old Harry having passed away or old Jimmy packing up because of his bad back. Here are sources of cheap equipment. Not necessarily good equipment because old beekeepers are noted for putting up with much loved and familiar equipment long after it really should have been changed.

Getting bees is relatively simple. Set out a hive with some used comb in it and wait for a swarm to take up residence. Success largely depends on how far away you are from the nearest beekeeper and could take some time or even fail altogether. A more certain way is to inform local police officers and pest control officers, both of whom are told of swarms having landed in a variety of odd spots, that you are prepared to collect a swarm within a given distance of your home. You should undoubtedly get some bees that way. Do have a care to check before your journey that they are actually a swarm of bees and not an underground bumble bee nest.

We now look at the management of the bees. It is a fact that the less bees are disturbed by the beekeeper the better they are for it and the more honey you will be able to gather. Disease in bees has become an ever-increasing problem over recent years and must be addressed at all costs. Gone are the days when a coarse beekeeper need only take the roof off a hive twice a year. Once in the Spring to check that the bees flying in and out are actually living there and not robbing and to put some supers on and again in late summer to take off the honey supers. Unless disease is tackled there is little doubt that you will lose your bees. There is of course the short term option of requesting the seasonal bees officer visit you to check your bees. I say "short term" because success in any case depends on what you tell him and I fancy the man will soon whittle out the over-coarse beekeeper who is merely using him so learn quickly from him what you will need to do. The "term" gets very short if you try the old trick of "while you're in there could you mark and or clip the queen for me, add or remove supers" etc?

Otherwise management is mainly concerned with swarm prevention, queen rearing and honey harvesting. Swarm control means far too many visits to and manipulations of the hive and the colony or fiddling about with multi-gated boards (Snelgrove) to suit the true coarse beekeeper. If you allow the bees to swarm in their own time you can save all that work. This also has the effect that you may well be able to collect the resulting swarm from where it rests and put it into another of the late Harry's ?? hives. You will also get a new queen in your existing hive without the bother of all that troublesome queen rearing.

This leaves only the honey harvesting. Although it may be unusual advice for the coarse beekeeper a certain amount of time spent in the preparation will in the long run save both time and money. Buy unwired wax for your honey supers; it is cheaper. Cut sheets lengthwise into 4 equal strips and fit one strip at the top of each frame. Only the most profligate beekeeper would use more. The bees will form their own cells along and below these strips. When it comes to harvesting the honey remove the frames, cut carefully along the joint where the bee-made cells meet the provided foundation. Cut the oblong block of honey-filled comb into sizes to fit cut comb containers or old margarine tubs depending on the destination of the honey. Properly labelled cut-comb containers can be sold. That in old margarine tubs can be used to pay any tradesmen prepared to barter his labour for your honey. They are out there; I have had roofs mended and cars repaired.

The coarse beekeeper's preparation of the bees for winter is to go indoors and forget about them until spring.

There is no need to mention mouse guards because unless the late Harry had them fitted to the hives when he died the coarse beekeeper is unlikely to own any.

Similarly wasted is the advice not to brush any snow off the hives because it helps to insulate the colony. It would never cross the coarse beekeeper's mind to do such a thing.

And so the coarse beekeeper's year ends. If the advice on the unavoidable disease control has been followed the bees should survive the winter. They have after all survived several million of them without the ministrations of "proper beekeepers".

A COUNTRYMAN'S TALE

Bill Clark, Cambridge

Can animals and birds recognise human faces?

I was 12 years old, and for some time - with my father's supervision - had been using a 'four-ten' shotgun, but on this day I had been allowed to go by myself. I was at the edge of a wood, sitting in a 'hide' built in a ditch to shoot pigeons raiding the tares out in the field. I had just shot one, and set it up to lure more in close, when a crow came circling round, keeping well out of range and cawing continuously. After about an hour of this I gave up. Dad said, 'the old devils must have a nest nearby. We will sort them out tomorrow.' He blamed carrion crows for stealing his hens eggs, pecking out the eyes of new-born lambs, and with his game-keeping hat on, blamed them for just about every loss he hadn't got an answer for. His knowledge of the bird consisted of: 'it is useless to fire a shotgun through their nests, for besides a thick base of twigs they always line it with leaves and a foot (30cm) of sheep's wool, they also have the guile and cunning of a fox, better eyesight and they can count'!

Before 'sun-up' the following morning, he placed half a rabbit in range of the hide, and we both sat inside, dad with his '12 bore' at the ready. Finally it was time for milking, and surmising that the crows had seen us arrive, he made a great ploy of holding his gun in view and a long over-coat at arms length - hoping to look like both of us leaving - and pausing only to hiss, 'Don't cough or so much as blink an eye', he left me to it. Hardly daring to breath I had aimed my four-ten towards the rabbit for at least a further ten minutes, when I was startled by the sudden shake of my leafy roof and looked up to see a crow's bright beady eyes looking straight into mine! It had flown in from behind and now sped off in the same direction cawing loudly. Crows certainly know an empty coat, even from a distance!

I was 19 when I met my wife-to-be, a gentle town-bred girl, who I soon realised had an affinity for birds and animals, domesticated or wild, especially the injured, sick or just plain hungry; she could gain their trust and confidence in the blink of an eye, whereas I, a brash country-bred lad, was brought up to regard all animals and birds as here for our benefit or at worst pests to be eradicated - the family's pet cats and dogs were only with us to earn their keep in that role! Wendy and I made a good team from the start. We knew all our farm animals by name, they knew us and we certainly noticed immediately if one looked doleful or sickly.

One day in the late 1960s, I arrived home for lunch, to be informed that a starving kestrel was flying around the house. I smiled, and remarked that it was most likely an escaped tame one. Over the following days it would fly down to snatch pieces of meat from Wendy's hand, but whizz off to a distant tree the moment I put in an appearance. Through binoculars I ascertained that it had a wing feather sticking upright from an upper wing surface, and I surmised that it was starving. A kestrel hawk is not going to catch any food when diving with a whistling wing!

Kestrel with an injured wing.

Kestrel restored to health and released into the wild.

Over the years - fuelled by the fact that I had abandoned farming and moved into conservation - there were many instances of injured or abandoned creatures being passed into Wendy's care. But it was when I started leading school groups around the countryside, that I began to take notice of just how much an animal can recognise about us. I used sheep to graze our flower meadows, and with the aid of an electric fence, moved them from place to place. Well used to this regularity, a bleating stampede would occur whenever I appeared, each animal hoping to be first into the new plot. I turned this into a party trick. As long as I kept in the rear of the group of children and teacher/parents - preferably behind a father - the sheep would briefly peruse the group from afar, and carry on grazing. I would quietly ask the children, 'Would you like the sheep to come and say hello?' And then step forward to call, 'Come and say hello to the children!' However, occasionally I was picked out regardless, usually by the same sheep, and if I happened to be in the company of an adult group or working nearby with some volunteers, those sheep would be the ones to come as close as possible and actually look at faces, latching onto mine to give a bleat of recognition - when immediately the whole bleating flock would dash over.

I used to mention, during my nature talks, that when snow was on the ground the wild birds quickly learned to follow me and my wheelbarrow - no surprise there though, for any wild bird can recognise food from afar, and will listen out for other birds 'chatter' too. Even when wild birds excitedly followed me as I towed a trailer-load of nest boxes, it was only that they were following the boxes, their perspicacity sharpened by heavily gravid females, bereaved by the loss of 1500 large old trees in the previous gale. The fact that many of them were making their choices before I had even picked the box from the trailer, or stepped off the ladder, was the big surprise! However, the following April after another winter gale clearance, my credibility was stretched to the limit, though I must confess it was at first due to Wendy. 'What a lovely day! Just listen to all the bird song', she remarked as she met up with me patrolling during a busy Saturday. 'I think these two above us now are singing especially for us.' I paused, 'That's not song! They are scolding us.' The birds kept pace with us before dropping back as we left their 'territory'. I always carried binoculars whilst on patrol, and so later as Wendy headed homewards I stayed to watch from nearby. A family walked into the area and the birds flew towards them, but with hardly a twitter flew away again, next a single man traversed our previous route, again the birds only gave him the briefest investigation, and after an elderly couple had also passed through without causing any scolding I deemed it was time to test my preposterous theory.

Could wild birds such as Great Tits - *Parus major* - actually recognize an individual human? Yes, I know I am a bit set in my dress so perhaps I am easy to pick out from the crowd, but by little birds looking down from above! I walked back into the area and the two birds plunged into view, keeping pace with me exactly as they twittered and twittered, only to fall silent as I left their territory. I returned, and back they came, getting extremely excited when I stopped in the clearing at the centre of their patch. I tried to recollect exactly the tree that we had cleared away - I was sure there had been no box on it. Then I remembered; it had sliced down the side of a nearby tree breaking off two branches as it fell. Through my binoculars I could see two nail holes where a box had been, I followed it's probable trajectory and discovered it in the heart of a blackberry bush. In no time at all I was back with ladder and tools, and with two birds excitedly urging me on, fished out the box and nailed it back in place. This time it was no surprise when the male bird landed to inspect my handy-work before I had reached the bottom of the ladder!

Our garden has several nesting boxes for birds. Here a blue tit is at the entrance to its adopted home.

Wendy and 'her birds' are still giving me food for thought. Amongst the birds that expect her to throw out a few tidbits each day (there was once a duck that waddled hundreds of metres through the woods to tap on our back door with its beak) is one large carrion crow. No matter what she is dressed in, the moment she appears he gives a loud caw and floats in to take first pick almost at her feet. If she goes out in a hat on a cold day, or with an umbrella on a wet day, she sometimes has to give a call first, when he will return her call, and may also circle first; go out with a new coat, hat or umbrella, and he will perch much closer to survey the situation before giving his 'caw' and dropping down. If I go out with her, he stays high in a tree two hundred metres distance until I remove myself from the scene. On the few occasions that I have been charged with the food delivery he flies silently to his look-out tree, only floating silently in to the food when I am some distance away. It goes without saying that most of the other birds stay away until he has vouched for the safety! This year for the first time - I believe he is a second generation 'Wendy crow' - he and his mate have built their nest close by. Although high in a Cedar of Lebanon, they both still visit it 'crow fashion', flying by a round-a-bout route through the trees and then quickly slipping in at the rear. I can walk under the tree, I can sit under the tree, but look directly up the tree and a dark shape immediately floats off almost unseen through the branches! What my father would have said I shudder to think, and perhaps the crows are wondering just how much of his blood still courses through my veins!

Any old how, I have no sooner settled to my own satisfaction that birds and animals can recognise us as individuals, than I see another conundrum - or I should say hear one! Although my hearing isn't what it was, all this spring I have commented on the song of a bird near our abode with an extra loud, sweet and continuous song: it's the first thing I hear on waking and the last sound I hear before I go to sleep. I was sure this could only be a Mistle Thrush (*Turdus viscivorus*). However, I was very surprised to see it was a Blackbird (*Turdus merula*) when I at last caught a sight of it. I wondered if it was using some Mistle Thrush mimicry, but credited it to my poor hearing.

Blackbird *(Turdus merula)*; Baikonur-town, Kazakhstan, Yuriy Danilevsky.

Mistle Thrush *(Turdus viscivorus)*; Malene Thyssen, http://commons.wikimedia.org/wiki/User:Malene

Recently, whilst attending my one beehive, I listened to the Blackbird that is in possession of the next territory to 'Mistle Boy' - who I could still hear in the distance: This was certainly a very ordinary, and broken, blackbird melody, and the thought crossed my mind that he must be very jealous of his neighbour. There was also another 'strange' bird-song coming from the same tree; this consisted of a long descending note followed by two short notes, sung twice with a long break, I felt I had heard it before, but couldn't place the bird. I blame my bees, my mind was really on them, because it was a further ten minutes before I realised that the strange song never overlapped the Blackbird song, and it was at least another five before I remembered that a police siren sometimes emits those notes before going quiet. Then as I came to this conclusion, the bird changed its position to perch in full view, and proceeded to also splice in short bursts of an ambulance siren! Lying in bed listening to 'Mistle Boy' the following morning, I was wondering just where all this escalation will end, when it occurred to me that it has already! Both birds are still singing 24/7 in mid June - neither has stopped to feed any young 'cos no sensible 'blackbird' female has been coerced into mating with them. Next spring things will be back to normal.

I now have the time to follow up on some old beekeeping beliefs. An old fellow I knew in the 1940s was certain that bees recognise their keepers - 'Just see how they come to the entrance to look at us.' And also at that time, another old chap was sure his bees knew him by his smell and he would place an article of his own underclothing under the roof of any new colony before he looked through them - he would be pleased to know that modern tutors usually tell beginners: 'it is unwise to use deodorant before visiting your bees.' He also recommended putting a piece of 'wedding' or 'christening' cake inside the hive when telling the bees there has been a wedding or a birth in the beekeeper's family. Ah well; perhaps some beliefs are better left in the past!

NB
B

THE SACRED BEE IN RENNES-LE-CHÂTEAU MYTHOLOGY

Andrew Gough

The study of Rennes-Le-Château incorporates more facets of esoteric research than almost any other mystery, subjects such as suppressed heretical knowledge, sacred geometry, art history, codes, Mary Magdalene and Jesus Christ, the Holy Grail, Royal Families, lost lineages, Knights Templar, Cathars, Visigoths, and the Priory of Sion, to name a few. But there is another, lesser known facet that we will now explore; the diminutive insect known as the 'sacred bee'.

The 21st century has seen a revival of interest in bees, due sadly, to their apparent demise and the impact that this potential catastrophe might have on the world economy. Quite simply, the loss of earth's most industrious pollinator of plants and trees and producer of medicinal and health food products, such as honey, would result in an unprecedented economic disaster, as evidenced by the fact that every third spoonful of food we put in our mouth has been made possible by bees.

Prehistory is full of clues that hint at ancient man's fascination with bees, and so is the mystery of Rennes-Le-Château. However, to appreciate them we must first turn our attention to the genesis of bee adoration, back in the age when temples of Venus graced the Aude valley, not churches dedicated to Mary Magdalene. In the *Cave of the Spider*, near Valencia, Spain, a 15,000-year-old painting depicts a determined-looking figure risking his life to extract honey from a precarious cliff-side beehive. *Honey hunting* represents one of man's earliest hunter / gatherer pursuits – its very difficulty hinting at the genesis of the adoration of the bee in prehistory. And of course, it was the bee that led ancient shamans to the plants whose hallucinogens transported their consciousness into the spirit world of the gods. Curiously, recent research has revealed that the sound of a bee's hum has been observed during moments of state changes in consciousness, including individuals who have experienced alleged UFO abductions, apparitions, and near death experiences. Was this phenomenon known by the ancients and believed to have been one of the elements that made the bee special?

In Anatolia, a 10,000-year-old statue of the Mother Goddess adorned in a yellow and orange beehive-style tiara has led scholars to conclude that the Mother Goddess evolved into the Queen Bee around this time. At the Neolithic settlement of Catal Huyuk, rudimentary images of bees dating to 6540 BC form a circle above the head of a goddess figure, creating the first ever 'halo', while beehive inspired designs are stylistically portrayed on the walls of its most sacred temples. Not surprisingly, it was the Sumerians who soon emerged as the forefathers of organized beekeeping, known as Apiculture, and invented *Apitherapy*, or the medical use of bee products such as honey, pollen, royal jelly and venom.

Sumerian reliefs depicting the adoration of extraordinary winged figures have often been interpreted by alternative history writers as proof of extraterrestrial intervention. However, in the context of the beekeeping, it appears they simply portray the veneration of bees. Significantly, the images gave rise to the dancing goddess motif; a female dancer with her arms arched over her head that scholars identify as a bee goddess; a type of shamanic priestess. The motif, which would become central to Egyptian symbolism, appears to allude to the bees' unique ability to communicate through dance; the waggle dance as it is known, or the ability to locate food up to three miles from the hive and communicate its whereabouts through dance, a sort of prehistoric satellite navigation.

The adoption of bee symbolism in Egyptian society developed rapidly, and by the start of the First Dynasty Egypt was known as the 'Land of the Bee', and the Pharaoh carried the title 'Beekeeper', with a bee prominently displayed in his cartouche. Additionally, Egyptians used honey as an offering to the gods in the afterlife, as well as in the mummification process. In fact, the gold and black horizontal stripes on the desk mask of Tutankhamen, and other Egyptian regalia, reference the bee's similarly striped body. Clearly, the seed of bee veneration that was sown by the Sumerians had been harvested by the Egyptians.

Minoan and Greek mythology soon followed, and adopted the sacred bee as a vital element of their society, depicting bees on the statues of their most important gods and goddesses. They also developed the coveted position of female bee shamans, called *Melissas*, which later evolved into priestesses known as *Sybils*. On the other side of the globe, Mayan culture venerated the bee and depicted gods in its image in their most sacred temples, and bee-hut styled structures sprang up from Africa to Ireland.

The early Catholic Church adopted the bee as a symbol of the Popes authority; evidence of which can be seen in Vatican City in the beehive-inspired papal headdress of past popes. Political movements, such as Communism, drew upon the altruistic, drone-like, 'proletariat' behavior exhibited in beehives, as a blueprint for their ideologies. An example of the bee's appreciation by those formulating and influencing the politics of the day, is the *Order of the Illuminati*,a 'secret' society founded by the German philosopher Johann Adam Weishaupt on 1 May, 1776 – Labor Day in modern times – the day of the worker, or drone. Amazingly, Weishaupt had considered naming his order 'Bees' – not 'Order of the Illuminati'.

Not surprisingly, the bee was also an important symbol in Freemasonry, and was depicted in many Masonic drawings of the 18th and 19th centuries. At the heart of the Masonic tradition is the concept of industry and stability. The theme stems from the stable, regular and orderly society that is observed in a beehive. In Freemasonry, the beehive represents all that is proper in society, and could arguably be regarded as its most important symbol.

French Freemasonry soon spread to the United States of America, aided by early American forefathers such as Thomas Jefferson, who wrote passionately about the importance of bees, whilst President George Washington featured the beehive on his Masonic apron. In no uncertain terms, early American society borrowed many of its philosophical principals from Freemasonry. In fact, not only was the entire Western Region of the United States originally named after the bee (Deseret) but America's most iconic statue – Washington's Monument – pays homage to the insect in an astonishing way; it contains an inscription that recalls the importance of bees throughout history, it states: *"Holiness to the Lord Deseret"*, meaning *'Holiness to the Lord, the Honeybee.'*

Genesis of the bee's adoration in France

Our abridged history of bee adoration slowly moves into the realm of Rennes-Le-Château with Napoleon Bonaparte, the military and political leader of France who in the early nineteenth century revived his country's fascination with bees. The bee was a hugely important icon of Napoleon's reign, and his obsession with its symbolism gave rise to his nickname; *The Bee*. Napoleon would have grown up with the symbolism of the bee engrained in his psyche, for his homeland, Corsica, was required to pay the

Romans an annual tax in beeswax which was the equivalent of £200,000. The young emperor ensured that the bee was widely adopted in his court as well as on clothing, draperies, carpets and furniture all across France. By choosing the bee as the emblem of his reign, Napoleon was paying homage to Childeric (436 – 481), one of the 'long-haired' Merovingian Kings of the region known as Gaul. When Childeric's tomb was uncovered in 1653, it was found to contain 300 golden jewels, styled in the image of a bee. And, of course, these are the same bees that Napoleon had affixed to his coronation robe. Sadly, of the 300 bees only two have survived.

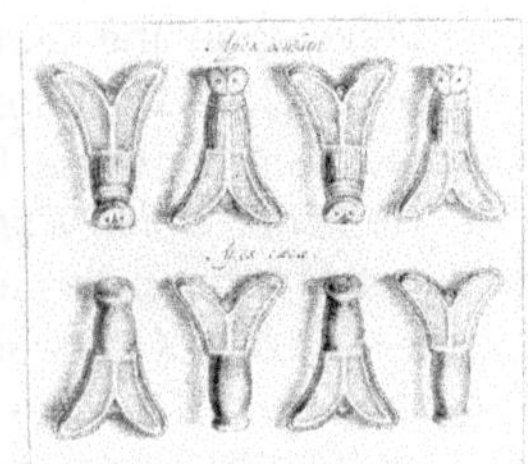

Bee's from the Tomb of Childeric I.

Childeric's hoard was entrusted to Leopold Wilhelm von Habsburg, a military governor of the Austrian Netherlands who was believed to have been a descendent of the Merovingian dynasty. Six years after his coronation, Napoleon married Marie-Louise, the daughter of Francis II, the last Habsburg to sit on the throne of the Holy Roman Empire. Napoleon's choice of the bee as the national emblem of his imperial rule spoke volumes about his desire to be associated with the Carolingians and Merovingian's; the early French kings whose funeral furniture featured bee and cicada symbolism as a metaphor for resurrection and immortality. The bee was also a vital symbol of French industry and one of the most prominent emblems of the French Revolution (1789–1799).

Across Europe, more than sixty cities selected an officially approved heraldry shield that included bees as part of its template. Remarkably, the bee was the precursor to the Fleur-de-lys; the national emblem of France. The theory is supported by many, including the French physician, antiquary and archaeologist, Jean-Jacques Chifflet. In fact Louis XII, the 35th King of France, was known as 'the father of the pope' and featured a beehive in his Coat of Arms. Disappointingly, his efforts to have the bee adopted as the Republic's official emblem were rejected by the National Convention due to their belief that *"Bees have Queens"*. Nevertheless, the bee remained a prominent element of French culture throughout the First and Second Empire (1804 to 1814, and 1852-1870) due to the enthusiastic patronage it had previously received.

Robert Lawlor studied the design of the bee and Fleur-de-lys in his book; 'Sacred Geometry' and concluded that the 1:√ proportion of the design is also found in the Islamic Mosque. Intriguingly, the mystical dimension of Islam known as Sufism maintained a secret brotherhood called Sarmoung, or Sarman, meaning bee. Members of the organization viewed their role as collecting the precious 'honey' of wisdom and preserving it for future generations. This (fascination with bees, and preserving wisdom for further generations) is precisely what connects us to Rennes-Le-Château, the unassuming but sombre hilltop hamlet in the shadow of the French Pyrenees. Here, at the turn of the 20th century, a group of priests – most famously Berenger Saunière, aroused suspicion with their curious behaviour and apparent wealth, leading many to speculate that they had discovered a great heretical secret – possibly involving Mary Magdalene, Jesus Christ, the Treasure of Solomon, hoards of the Visigoth's, or valuables hidden during the French Revolution. In reality, what they found, if anything, remains a mystery.

Although the legend of Rennes-Le-Château has struck a chord with modern day audiences, its roots stem from the Merovingian kings so revered by Napoleon, and its origins, ingrained in the psyche of so many of us, go like this: Childeric I fathered Clovis I, who succeed his father in 481 as king of the region that now borders Belgium and France, and in the process became the first ruler to unite the previously hostile and independent Frankish tribes. A line of descendants leads to Dagobert I, king of the Franks from 629–634, who fathered Sigelbert III, who fathered Dagobert II, who married Giselle de Razes, the daughter of the Count of Razes and the niece of the king of the Visigoths. The two were said to have married at Rhedae, a stronghold widely believed to be Rennes-Le-Château, although the association remains unconfirmed. Years later, in 754 AD, Childeric III died childless, marking the end of a dynasty that had been in decline since Dagobert II was assassinated near Stenay-sur-Meuse on December 23rd, 679 AD.

Many believe that the *Dalle des Chevaliers*, or Knights Stone as it is known today, recalls a portion of this history. The stone portrays two scenes, each consistent with the Carolingian style of the eight century, and the popular interpretation is that the primary scene depicts baby Sigebert being carried by a horseman to his mother in Rennes-Le-Château, while some speculate that the Knight is carrying the Holy Grail. Curiously, Wolfram von Eschenbach, who wrote history, never fiction, compiled the first complete Grail Romance, *Parzival*, and in his account we are told that the Grail is a stone from heaven. This is interesting, given that the word 'meteorite' carries the same numeric value (443) in the Cabala as 'Bethel' – which translates as 'bee' in Egyptian, and many believe the grail to be an oracle stone that fell from the heavens in antiquity, most likely in Egypt.

The belief that the Merovingians were special, and that they represented a royal bloodline, led Napoleon to commission an extensive analysis of their lineage. Fascination with the mysterious line of kings continued into the 20th century when a Frenchman by the name of Louis Vazart founded an organisation based in Stenay-sur-Meuse called 'Cercle Saint Dagobert II', that specialised in the study of Merovingians, and Dagobert II in particular. For its logo, Vazart chose an image of a bee inside of a 6-sided cone, or Hexagon – the shape of a beehive cell, surrounded by a circle.

Logo of Cercle Saint Dagobert II; A bee in a Hexagon.

Vazart's selection of the bee is consistent with his research, for France itself is known as *l'Hexagone*, due to its natural 6-sided shape. Coincidently, the centre line of *l'Hexagone* closely mirrors the old Paris Meridian, passing near Paris in the north and Rennes-le-Chateau in the south. The Paris Meridian – an imaginary arc that measures the hours of the day, was later replaced by London's Greenwich Meridian as the international standard for time keeping. However, in recent years, the Paris Meridian has been romanticised and somewhat merged with the notion of the Rose-Line, a mythical sort of ley-line connecting esoterically significant sites from Roslyn Chapel in Scotland to Saint Sulpice in Paris, and on to Rennes-Le-Château in the south of France. Despite its questionable authenticity, it is worth mentioning that the two sites that top and tail the Rose-Line, Roslyn Chapel and Rennes-Le-Château, each feature bee symbolism, and in peculiar ways.

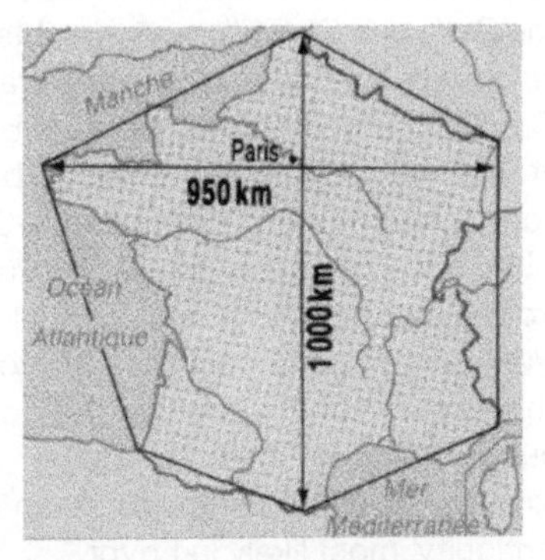

France – believed to be in the shape of a natural Hexagon

Before delving further into Roslyn Chapel, it is worth mentioning that in a similar vein to the *l'Hexagone* symbolism, Philippe de Cherisey, a friend of Plantard's who founded a magazine called Circuit, whose distribution was said to include the membership of the Priory of Sion, featured a hexagon imprinted over an image of France with a sword symbolically piercing its centre, echoing the old Paris Meridian.

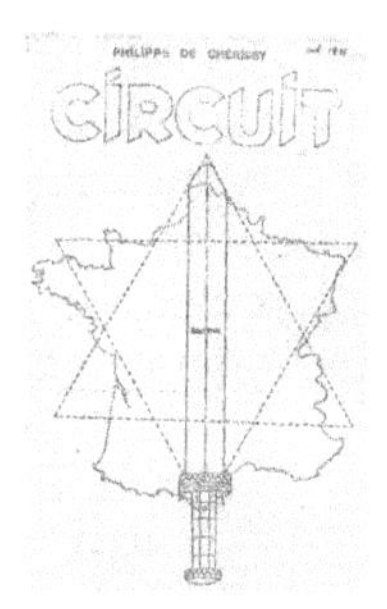

The cover of Philippe De Cherisey's Circut

Roslyn Chapel was founded by William Sinclair, 1st Earl of Caithness, in the 15th century and is renowned for what many believe to be an elaborate display of Masonic symbolism. In fact, some believe that the chapel contains treasures of the Knights Templar or even the Holy Grail itself. Hyperbole aside, Roslyn Chapel does in fact contain a splendidly carved column known as the Apprentice Pillar, or *Princes Pillar* as it was called in ancient accounts. The pillar, which stands to the right of the church alter, is adorned with what is generally regarded as *Tree of Life* symbolism; two dragons of Yggdrasil – the World Tree according to Norse Mythology – reside at its base, and a masonry vine spirals vertically around the column, drawing our attention to the ceiling.

The Princes Pillar – Roslyn Chapel, Scotland

Recent theories put forth by Alan Butler and John Ritchie in their book; *'Rosslyn Revealed: A Library in Stone'*, suggest that the ceiling above the Princes Pillar represents "paradise" on earth. And on the roof of the chapel we find a curious stone beehive with a lone flower peddle entrance that was homed by bees for as long as anyone can remember. Sadly, the bees were removed in the 1990's and have not returned. However, the existence of the beehive in the proximity of the vine recalls a biblical account of a staff that grows into a great tree, with; *"a vine twisted around it and honey coming from above."* Might the design of the roof, ceiling and Princes Pillar, reflect the role of bees and honey in the greater context of paradise and the World Tree of Life? Curiously, the association of the Tree of Life with hexagonal beehive symbolism is not unique. In fact, it is featured on the new Euro coin, reinforcing the importance of the symbolism to this day.

The new Euro Coin: Tree of Life Symbolism

From Roslyn Chapel in the north, the mythical Rose-Line reunites with Rennes-le-Chateau in the south, the village with alleged Merovingian connections. Although history informs us that the Merovingian dynasty died out with Dagobert II, this has not prevented others from claiming descent, such as Pierre Plantard, a Frenchman who in the 20th century promoted his association with the Merovingians, as well as Rennes-le-Chateau, and was regarded by some as the last direct descendant of Jesus Christ. Plantard also claimed to have been a Grand Master of the Priory of Sion, a controversial society with considerable interests in the Merovingian lineages commissioned by Napoleon. Curiously, Plantard's family crest featured both the Fleur-de-lys and eleven bees.

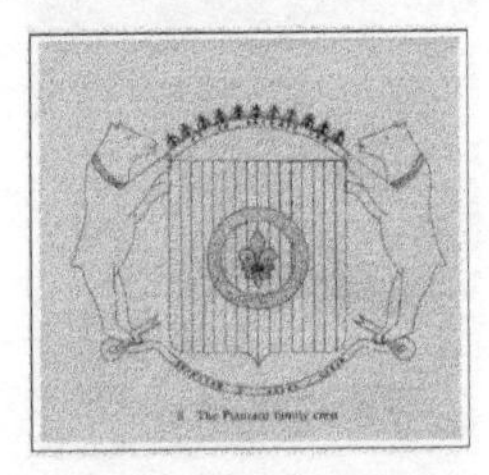

Plantard Family Crest

Rennes-Le-Château and the Bee

Rennes-Le-Château is linked with bees in curious ways, some complex, others just peculiar, such as the experience of Christopher Dawes, author of the Rennes-Le-Château adventure yarn; *'Rat Scabies and the Holy Grail'*, who inexplicably encountered dead bees while conducting his research. Another is found in *'The Key to the Sacred Pattern'* by Henry Lincoln, co-author of the 1982 book; *'Holy Blood and the Holy Grail'*, the international bestseller that put Rennes-Le-Château on the map with English speaking audiences around the world. Here, Lincoln draws our attention to a series of beehive-like huts called "Capitelles" found near a village called Coustaussa – site of a macabre assignation of a priest, Abbe Gelis, a friend of Saunière's who appears to have become dangerously entwined in the mystery. The huts, which are largely unexcavated, are part of what is known locally as the "Great Camp". The beehive-like structures mirror others across the globe, and are one of the few artefacts that lend credence to the belief that Rennes-Le-Château may have been the ancient Visigoth settlement of Rhedae.

Yet another link stems from the infamous Latin expression that hangs over the door of the village church of Saint Mary Magdalene in Rennes-Le-Château; 'TERRIBILIS EST LOCUS ISTE', meaning *This Place Is Terrible*. The biblical phrase refers to the words that Jacob spoke when he awoke from his dream about a ladder that reached to heaven. To this end, Genesis 35:1 provides the reference; *"And God said unto **Jacob**, Arise, go up to **Bethel**, and dwell there: and make there an altar unto God."* So Jacob recounted that the place was called Bethel and had a stone erected commemorating the spot where he had fallen asleep. The biblical story relates to the bee in that Bethel, or *Bytal* in Hebrew, means 'House of God', and the letter 'Y' and the letter 'I' are interchangeable, rendering the translation 'Bit-al', and 'Bit' in ancient Egyptian means bee. The translation also suggests that *House of God* may represent a repository of knowledge – as in the beehive. Additionally, Bethel carries the same numeric value as 'meteorite', which harkens back to the notion that bees are related to sacred stones from heaven.

After reviewing several rather obtuse links between bees and Rennes-Le-Château, there is another association that stands apart from the rest in its implications, and it involves Henry Lincoln and the french author Gerard de Sede, whose 1967 book, *'The Accursed Treasure of Rennes-Le-Château'*, first catapulted the mystery to prominence in France. The story goes that Lincoln purchased de Sède's book whilst on holiday in France and succeeded in deciphering one of its peculiar parchments, giving spark to the flame that still burns today; what if anything do the coded parchments conceal? Lincoln later came across a 'Book Club' version with a strange photograph of bees not referenced in the text. Incongruously, the title beneath the photo simply stated, *'Rennes-les-Bains – Thermes Romains'*, and no other reference to the photograph was made.

The anomaly is recounted in Lincoln's 1998 book, *'The Key to the Sacred Pattern'*. Essentially, the photo depicts a wooden panel on a dining room door with four bees, one in each corner, and in the middle, a winged female standing on a globe holding a wreath above her head like an Egyptian dancing goddess – a motif we now understand to represent bee goddesses, Melissas and Sybils. Later, de Sède provided Lincoln with material for his BBC television special about Rennes-Le-Château, including photos taken by Plantard that de Sède had used in his book.

In *'The Key to the Sacred Pattern'*, Lincoln recounts how the back of the photos were stamped with a seal saying "PLANTARD", along with an explanation of how the woman in the centre of the photograph was Europa, the legendary priestess who was seduced by Zeus while he was in the form of a bull ('Apis,' the most sacred bull in antiquity, is Latin for 'bee'), and that the accompanying images of bees represented apiculture. Fair enough, but what is most intriguing is that the notation on the back of the photographs is said to have included the phrase, "We are the Beekeepers", a detail not revealed by Lincoln in his book. The expression recalls the 'Beekeeper' title held by Egyptian Pharaohs and begs the question, was Plantard inferring that he was a Beekeeper – and if so, of what – the Priory of Sion?

Clearly, ancient societies, including France and the forefathers of the Rennes-Le-Château mystery, believed that bee symbolism was quite important. They may have even considered themselves beekeepers, but of what exactly, remains to be determined. Not surprisingly, the existence of the sacred bee has all but faded from memory; the question remains, however, is that what the beekeepers intended?

BOOKS: BEATING THE LIBRARY FUNDING SHORTAGES

John Kinross

"Principles of Bee Improvement" Jo Widdicombe (NBB £11.95). This is the first of my choices of this year's books. Jo is a Cornish bee farmer. He starts by saying "we produce about 1/6th of the honey we consume" in the UK; he is determined to to do something to increase the shortfall. However, Cornwall is not the best county for beekeeping. Shelter is scarce and everywhere is too close to the sea, so sometimes you are in a sea mist for days.

He discusses different types of queen rearing, recommending the "Queen-right System" by Brown and Wilkinson (more details here would have been welcome, Jo). I note that he uses a poultry egg incubator. I still remember the ghastly smell of them. Perhaps Jo has no problem with this when used for bees.

It is a well-put-together book which will be worth having in your library.

Margaret Cowley's "The Honey Bee" (Bee Craft £12.50) is a special copy of the magazine comprising clear anatomical pictures of the bee for those studying anatomy. Another well-put-together book, but it has no index. The student will have no problem copying the diagrams into a note book. The information is certainly easier to read than the material found in Dade or Snodgrass, though I still prefer the BBKA Leaflet on Bee Anatomy (for left-handed beekeepers) which is now long out of print.

Wally Shaw's "An Apiary Guide to Swarm Control" (NBB £7.95) is a useful paper on the subject for beekeeping groups. I would have liked a glossary, bibliography and index, but this is a reprint of an article and is good value for its price. Wally keeps his bees in Anglesey and is a stalwart of the Welsh BKA.

There is also **Tony Jefferson's "A Practical Guide to Producing Heather Honey"** (NBB £8.95), another slim publication. It has some excellent colour pictures, one a bit scary of a grass snake, and one can feel the wind when we see all those stones on the hive roofs. It must be difficult to find really sheltered places. The processing of heather honey is a difficult job, especially for the first time, but Tony seems to manage it all and explains it with a 'devil-may-care' attitude which is very refreshing.

There was one book out at the National Honey Show written by an old friend. Those of you who can remember shows at Paddington and other strange places in London will recall a bare table where one man cut up wood and knocked in nails all day without worrying if anything was finished or sold. **Norman Chapman**, the name of my friend, has now written **"Pollen Microscopy"** (CMI, 132 pp, £21), having used his cameras and ideas from Dorothy Hodge's famous book "Pollen Loads of the Honeybee" to make large pollen drawings to go with his collection of photos. The result is a book for all bee microscopists. A warning about lime trees though. Some are very poisonous and Norman needs to put down which lime he has taken a photograph of. I like his Mounting Press pad which seems to have been made of strawberry ice cream. Perhaps it is tea time? Well done, Norman! I can now give up trying to make an extractor out of a dustbin as described in your previous book - "DIY Beekeeping".

Finally, a book for all bee people who want to know what the bee they have just rescued (usually from our church) is called. **S Falk's "Field Guide to Bees"** illustrated by R Levington has 1000 illustrations, 700 in colour, 234 maps and is well worth £35 (Bloomsbury Press). Dave Goulson says it is "a long awaited and authoritative field guide".

I hope that you find the book that you want. Don't forget to remind your association librarian, if you have one, to get some of these books with his or her funds if you don't have room on your own bookshelves. We have no library in Hereford city now due to 'Asbestos' (or council money shortages), so some of us put on a Welsh accent and go to Monmouth where they couldn't be more helpful.

WORD SEARCH

ANATOMY OF THE HONEYBEE

The grid below contains thirty anatomical parts of the honeybees' bodies (worker, queen and drone). They may be found by searching the grid horizontally, vertically and diagonally, both backwards and forwards.

E A O I Q M Q A P R O B O S C I S E R H

D V V Q Z X U X O Y A R B U D E U W K U

H Y P O P H A R Y N G I A L L I S N E S

D G L H T E L D B Z I S I U R P Y W C Q

G N Q Z N F N Y E A Y P B C Z M C R X A

W O V M U L L E B A L U C I B R O C D B

V E N O M S A C E H T A M R E P S K I D

R Q S S E L B I D N A M J T A B E L Y O

R J D E S U L L A H P O D N E A L J H M

A S L T B O E I A C H C Z E H E C P T E

S P K A S C H E V N I K O V C P A S L N

T U K E I G M E S B N L K O A E R C G I

E Y G D I O B T A A O E W R R I I D N V

L I E P L J O R C E E X T P T R P A R U

L P L Y S N I R O Y L Y A N U M S M U L

U A M U E U H N X W B A V A A A M U P S

M P L F M S K O V F C H G H N K G X Y R

H S W E X A R O H T S P O O N T N A O P

I W F D Q M H J C Q A D V V M I X T M O

X N S K E Z B F D Z J U K P P B W Q S H

ANSWERS ON PAGE 194

17

DIARY & CALENDAR

- PART II -

*SR (SUNRISE) SS (SUNSET) FOR LONDON UK.

JANUARY

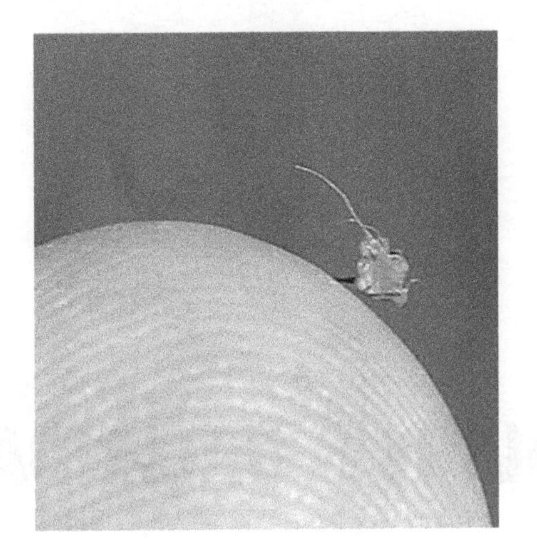

"He is not worthy of the honey-comb
That shuns the hives because the bees have stings."

William Shakespeare
April 26th 1564 (?) - April 23rd 1616

DAY	JANUARY 2017 FORAGE	TEMP		WIND		CL'D	RAIN	1	2	3
		MIN	MAX	DIR	B.S			HIVE WEIGHT		
1										
2										
3										
4										
5										
6										
7										
8										
9										
10										
11										
12										
13										
14										
15										
16										
17										
18										
19										
20										
21										
22										
23										
24										
25										
26										
27										
28										
29										
30										
31										

JAN17

1,SU
NEW YEAR'S DAY

2,MO
BANK HOLIDAY

3,TU

4,WE

5,TH
12TH NIGHT

6,FR
EPIPHANY

7,SA
SR:08:05, SS:04:08
ORTHODOX FEAST OF THE NATIVITY

8,SU

9,MO

10,TU

11,WE

12,TH ○

13,FR

14,SA
SR:08:00, SS:04:016

15,SU

16,MO	24,TU
17,TU	25,WE BURN'S NIGHT
18,WE	26,TH
19,TH	27,FR
20,FR	28,SA ● SR:07:45, SS:04:42
21,SA SR:07:54, SS:04:03	29,SU
22,SU	30,MO
23,MO	31,TU

FEBRUARY

"Those who have handled sciences have been either men of experiment or men of dogmas. The men of experiment are like the ant, they only collect and use; the reasoners resemble spiders, who make cobwebs out of their own substance. But the bee takes a middle course: it gathers its material from the flowers of the garden and of the field, but transforms and digests it by a power of its own. Not unlike this is the true business of philosophy; for it neither relies solely or chiefly on the powers of the mind, nor does it take the matter which it gathers from natural history and mechanical experiments and lay it up in the memory whole, as it finds it, but lays it up in the understanding altered and digested."

Francis Bacon, 1st Viscount St Alban,
22nd January 1561 – 9th April 1626

This quotation is also accredited to Leonardo da Vinci
Illustration from a painting by Francesco Melzi

DAY	FEBRUARY 2017 FORAGE	TEMP		WIND		CL'D	RAIN	1	2	3
		MIN	MAX	DIR	B.S			HIVE WEIGHT		
1										
2										
3										
4										
5										
6										
7										
8										
9										
10										
11										
12										
13										
14										
15										
16										
17										
18										
19										
20										
21										
22										
23										
24										
25										
26										
27										
28										

FEB17

	8,WE
1,WE	**9,TH** SHROVE TUESDAY, MARDI GRAS
2,TH ○ CANDLEMAS DAY	**10,FR** ASH WEDNESDAY
3,FR	11,SA ○ SR:07:22, SS:05:07 ST GOBNAIT, PATRON SAINT OF IRISH BEEKEEPERS
4,SA SR:07:34, SS:04:54	12,SU
5,SU	**13,MO** ST MODMONOC
6,MO	**14,TU** ST VALENTINE
7,TU	**15,WE**

16,TH

17,FR

18,SA
SR:07:09, SS:05:20

19,SU

20,MO

21,TU

22,WE

23,TH
ST KHALAMPII, PATRON SAINT OF BULGARIAN BEEKEEPERS (HIVE-SHAPED PIES BAKED)

24,FR

25,SA
SR:06:54, SS:05:33

26,SU ●

27,MO

28,TU
SHROVE TUESDAY

MARCH

"Went drearily singing the chore-girl small."

Telling the Bees

" - It was only for folks that died in the house, he said. But he had himself told the bees when his wife died. He had gone out on that vivid June morning to his hives, and had stood watching the lines of bees fetching water, their shadows going and coming on the clean white boards. Then he had stooped and said with a curious confidential indifference, 'Maray's jead.' He had put his ear to the hive and listened to the deep, solemn murmur within; but it was the murmur of the future, and not of the past, the preoccupation with life, not with death, that filled the pale galleries within. Today the eighteen hives lay under their winter covering, and the eager creatures within slept. Only one or two strayed sometimes to the early arabis, desultory and sad, driven home again by tile frosty air to await the purple times of honey. The happiest days of Abel's life were those when he sat like a bard before the seething hives and harped to the muffled roar of sound that came from within."

From - *"Gone to Earth"*

Mary Webb

March 25th 1881 – October 8th 1927

DAY	MARCH 2017 FORAGE	TEMP MIN	TEMP MAX	WIND DIR	WIND B.S	CL'D	RAIN	HIVE WEIGHT 1	HIVE WEIGHT 2	HIVE WEIGHT 3
1										
2										
3										
4										
5										
6										
7										
8										
9										
10										
11										
12										
13										
14										
15										
16										
17										
18										
19										
20										
21										
22										
23										
24										
25										
26										
27										
28										
29										
30										
31										

MAR17

	8,WE
1,WE ST DAVID'S DAY ASH WEDNESDAY	9,TH
2,TH ○	10,FR
3,FR	11,SA SR:06:24, SS:05:57
4,SA SR:06:39, SS:05:45	12,SU ○
5,SU	13,MO
6,MO	14,TU
7,TU	15,WE

16,TH	**24,FR**
17,FR ST PATRICK	25,SA SR:06:45, SS:07:26 GREEK INDEPENDENCE DAY
18,SA SR:06:08, SS:06:09	26,SU DST CLOCKS TURNED FORWARD ONE HOUR MOTHERING SUNDAY
19,SU	**27,MO**
20,MO SPRING EQUINOX - FIRST DAY OF SPRING, PALM SUNDAY	**28,TU** ●
21,TU	**29,WE**
22,WE	**30,TH** ST ALEXIUS (UKRAINIAN BEEKEEPERS HANG ICONS OF THEIR PATRON SAINTS OF BEEKEEPING, ST SAVVATY AND ST ZOSIMA IN SHRINES AMONGST THEIR HIVES)
23,TH	**31,FR**

APRIL

"The keeping of bees is like the direction of sunbeams."

Henry David Thoreau
July 12th, 1817 – May 6th, 1862

DAY	APRIL 2017	TEMP		WIND		CL'D	RAIN	1	2	3
	FORAGE	MIN	MAX	DIR	B.S			HIVE WEIGHT		
1										
2										
3										
4										
5										
6										
7										
8										
9										
10										
11										
12										
13										
14										
15										
16										
17										
18										
19										
20										
21										
22										
23										
24										
25										
26										
27										
28										
29										
30										

APR17

	8,SA SR:06:21, SS:07:44
1,SA SR:06:36, SS:07:32	9,SU PALM SUNDAY
2,SU	10,MO
3,MO	11,TU ○
4,TU	12,WE
5,WE	13,TH
6,TH	14,FR GOOD FRIDAY (AND ORTHODOX)
7,FR	15,SA SR:06:05, SS:07:56

16,SU APRIL EASTER SUNDAY (AND ORTHODOX)	24,MO
17,MO EASTER MONDAY (AND ORTHODOX)	25,TU
18,TU	26,WE ●
19,WE	27,TH
20,TH	28,FR
21,FR	29,SA SR:05:37, SS:08:19
22,SA SR:05:51, SS:08:07	30,SU ST ZOSIMA - 'GREET THE BEE ON ZOSIMA'S DAY AND THERE WILL BE HIVES AND WAX'.
23,SU ST GEORGE	

MAY

"I will arise and go now, and go to Innisfree,
And a small cabin build there, of clay and wattles made:
Nine bean-rows will I have there, a hive for the honey bee,
And live alone in the bee-loud glade."

From - *The Lake Isle of Innisfree*

William Butler Yeats
July 4th 1808 – April 25th 1879
Photograph by Alice Boughton in 1903

DAY	MAY 2017 FORAGE	TEMP		WIND		CL'D	RAIN	1	2	3
		MIN	MAX	DIR	B.S			HIVE WEIGHT		
1										
2										
3										
4										
5										
6										
7										
8										
9										
10										
11										
12										
13										
14										
15										
16										
17										
18										
19										
20										
21										
22										
23										
24										
25										
26										
27										
28										
29										
30										
31										

MAY17

	8,MO
1,MO BANK HOLIDAY, MAY DAY (EASTERN EUROPE)	9,TU
2,TU	10,WE ○
3,WE	11,TH
4,TH	12,FR
5,FR	13,SA SR:05:12, SS:08:41
6,SA SR:05:24, SS:08:30	14,SU
7,SU	15,MO

16,TU	24,WE
17,WE	25,TH ● ASCENSION DAY
18,TH	26,FR
19,FR	27,SA SR:04:54, SS:09:01
20,SA SR:05:02, SS:08:52	28,SU
21,SU	29,MO SPRING BANK HOLIDAY
22,MO	30,TU
23,TU	31,WE

JUNE

"The bee's life is like a magic well:
the more you draw from it, the more it fills with water".

From - *'Bees: Their Vision, Chemical Senses and Language'*

Karl Ritter von Frisch
November 20th 1886 – June 12th 1982
Nobel Prize Winner in 1973, along with Nikolaas Tinbergen and Konrad Lorenz

	JUNE 2017	TEMP		WIND		CL’D	RAIN	1	2	3
DAY	FORAGE	MIN	MAX	DIR	B.S			HIVE WEIGHT		
1										
2										
3										
4										
5										
6										
7										
8										
9										
10										
11										
12										
13										
14										
15										
16										
17										
18										
19										
20										
21										
22										
23										
24										
25										
26										
27										
28										
29										
30										

JUN17

1,TH

2,FR

3,SA
SR:04:46, SS:09:12

4,SU

5,MO

6,TU

7,WE

8,TH

9,FR ○

10,SA
SR:04:45, SS:09:15

11,SU

12,MO

13,TU

14,WE

15,TH

16,FR	24,SA ● SR:04:45, SS:09:20
17,SA SR:04:43, SS:09:19	25,SU
18,SU FATHER'S DAY	26,MO
19,MO	27,TU
20,TU	28,WE
21,WE SUMMER SOLSTICE - FIRST DAY OF SUMMER	29,TH
22,TH	30,FR
23,FR	

JULY

"We entered and looked at the cells which circled the courtyard.
The monk extended his hand. 'Behold God's bee hive', he said sarcastically,
'Behold the cells. Once they were inhabited by bees who made honey;
now by drones, and what a sting they have . . .
May the Lord protect you',
he added and burst out laughing."

From *"Report to Greco"* in which the author describes his month-long travels on Mount Athos and quotes words spoken to him by a monk at Agios Pavlos.

Nikos Kazantzakis
February 18th 1883 – October 26th 1957
Bust of N.K. above the beach of Kalogria, Stoupa, where he wrote "Zorba the Greek". Although the story was set in Crete, he was inspired by the lignite mining above this coastal resort.

DAY	JULY 2017 FORAGE	TEMP		WIND		CL'D	RAIN	1	2	3
		MIN	MAX	DIR	B.S			HIVE WEIGHT		
1										
2										
3										
4										
5										
6										
7										
8										
9										
10										
11										
12										
13										
14										
15										
16										
17										
18										
19										
20										
21										
22										
23										
24										
25										
26										
27										
28										
29										
30										
31										

JUL17

1,SA SR:04:49, SS:09:19	8,SA SR:04:54, SS:09:15
2,SU	9,SU ○
3,MO	10,MO
4,TU	11,TU
5,WE	12,WE
6,TH	13,TH
7,FR	14,FR ST SWITHUN
	15,SA SR:05:02, SS:09:19

16,SU	24,MO
17,MO	25,TU
18,TU	26,WE
19,WE	27,TH
20,TH	28,FR
21,FR	29,SA SR:05:21, SS:08:51
22,SA SR:05:11, SS:09:01	30,SU
23,SU ●	31,MO

AUGUST

"There's a whisper down the field where the year has shot her yield,
And the ricks stand grey to the sun,
Singing: 'Over then, come over, for the bee has quit the clover,
'And your English summer's done' ."

From *'The Long Trail'*

Joseph Rudyard Kipling
30th December 1865 – 18th January 1936

DAY	AUGUST 2017 / FORAGE	TEMP		WIND		CL'D	RAIN	HIVE WEIGHT		
		MIN	MAX	DIR	B.S			1	2	3
1										
2										
3										
4										
5										
6										
7										
8										
9										
10										
11										
12										
13										
14										
15										
16										
17										
18										
19										
20										
21										
22										
23										
24										
25										
26										
27										
28										
29										
30										
31										

AUG17

1,TU	8,TU
2,WE	9,WE
3,TH	10,TH
4,FR	11,FR
5,SA SR:05:31, SS:08:40	12,SA SR:05:43, SS:08:27
6,SU	13,SU
7,MO ○ SUMMER BANK HOLIDAY (SCOTLAND)	14,MO
	15,TU ORTHODOX DORMITION B V MARY

16,WE

17,TH

18,FR

19,SA
SR:05:53, SS:08:13

20,SU

21,MO

22,TU

23,WE

24,TH
ST BARTHOLOMEW
(TRADITIONAL DAY FOR HARVESTING HONEY)

25,FR

26,SA
SR:06:04, SS:07:58

27,SU

28,MO
SUMMER BANK HOLIDAY (ENG. WALES, NI)

29, TU ●

30, WE

31,TH

SEPTEMBER

"One can no more approach people without love than one can approach bees without care. Such is the quality of bees..."

Count Lev Nikolayevich Tolstoy
28th August 1828 – 20th November

DAY	SEPTEMBER 2017 FORAGE	TEMP		WIND		CL'D	RAIN	1	2	3
		MIN	MAX	DIR	B.S			HIVE WEIGHT		
1										
2										
3										
4										
5										
6										
7										
8										
9										
10										
11										
12										
13										
14										
15										
16										
17										
18										
19										
20										
21										
22										
23										
24										
25										
26										
27										
28										
29										
30										

SEP17

	8,FR
1,FR	9,SA SR:06:27, SS:07:27
2,SA SR:06:15, SS:07:43	10,SU
3,SU	11,MO
4,MO	12,TU
5,TU	13,WE
6,WE ○	14,TH
7,TH	15,FR

16,SA
SR:06:38, SS:07:11

17,SU

18,MO

19,TU

20,WE

21,TH

22,FR
AUTUMN EQUINOX - FIRST DAY OF AUTUMN

23,SA
SR:06:49, SS:07:11

24,SU

25,MO

26,TU

27,WE

28,TH ●

29,FR

30,SA
SR:07:00, SS:06:39

OCTOBER

"The little bee returns with evening's gloom,
To join her comrades in the braided hive,
Where, housed beside their might honey-comb,
They dream their polity shall long survive."

Charles Tennyson Turner
4th July 1808 – 25th April 1879

Born in Somersby, Lincolnshire, he was an elder brother of Alfred Tennyson

	OCTOBER 2017	TEMP		WIND		CL'D	RAIN	1	2	3
DAY	FORAGE	MIN	MAX	DIR	B.S			HIVE WEIGHT		
1										
2										
3										
4										
5										
6										
7										
8										
9										
10										
11										
12										
13										
14										
15										
16										
17										
18										
19										
20										
21										
22										
23										
24										
25										
26										
27										
28										
29										
30										
31										

OCT17

8,SU

1,SU

9,MO

2,MO

10,TU

3,TU

11,WE

4,WE
ST FRANCIS - ANIMAL DAY

12,TH

5,TH ○

13,FR

6,FR

14,SA
SR:07:23, SS:06:08

7,SA
SR:07:12, SS:06:23

15,SU

16,MO	**24,TU**
17,TU	**25,WE**
18,WE	**26,TH**
19,TH ●	**27,FR**
20,FR	28,SA SR:07:48, SS:05:29 OXI DAY (GREECE)
21,SA SR:07:35, SS:05:53	29,SU DST CLOCKS TURNED BACK ONE HOUR
22,SU	**30,MO**
23,MO	**31,TU** HALLOWEEN

NOVEMBER

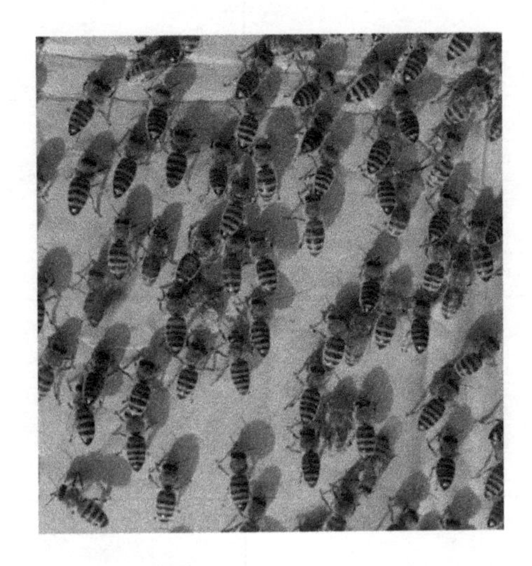

"With breezes from our oaken glades,
But thou wert nursed in some delicious land
Of lavish lights, and floating shades:
And flattering thy childish thought
The oriental fairy brought,
At the moment of thy birth,
From old well-heads of haunted rills,
And the hearts of purple hills,
And shadow'd coves on a sunny shore,
The choicest wealth of all the earth,
Jewel or shell, or starry ore,
To deck thy cradle, Eleanore.

Or the yellow-banded bees,
Thro' half-open lattices
Coming in the scented breeze,
Fed thee, a child, lying alone,
With whitest honey in fairy gardens cull'd —
A glorious child, dreaming alone,
In silk-soft folds, upon yielding down,
With the hum of swarming bees
Into dreamful slumber lull'd."

From - *"Eleaanor"*

Alfred Tennyson Tennyson
August 6th 1809 – October 6th 1892

DAY	NOVEMBER 2017 FORAGE	TEMP MIN	TEMP MAX	WIND DIR	WIND B.S	CL'D	RAIN	HIVE WEIGHT 1	HIVE WEIGHT 2	HIVE WEIGHT 3
1										
2										
3										
4										
5										
6										
7										
8										
9										
10										
11										
12										
13										
14										
15										
16										
17										
18										
19										
20										
21										
22										
23										
24										
25										
26										
27										
28										
29										
30										

NOV17

	8,WE
1,WE ALL SAINTS DAY	9,TH
2,TH ALL SOULS' DAY	10,FR
3,FR	11,SA SR:07:12, SS:04:15
4,SA ○ SR:07:00, SS:04:27	12,SU REMEMBRANCE DAY
5,SU GUY FAWKES DAY	13,MO
6,MO	14,TU
7,TU	15,WE

16,TH	**24,FR**
17,FR POLYTECHNEIO (GREECE)	25,SA SR:07:35, SS:03:58
18,SA SR:07:24, SS:04:06	26,SU ●
19,SU	**27,MO**
20,MO	**28,TU**
21,TU	**29,WE**
22,WE	**30,TH** ST ANDREW
23,TH THANKSGIVING DAY (USA)	

DECEMBER

"Stands the Church clock at ten to three?
And is there honey still for tea?"

From *"The Old Vicarage, Grantchester"* 1912

Rupert Brooke
August 3rd 1887 – April 23rd 1915

	DECEMBER 2017	TEMP		WIND		CL'D	RAIN	1	2	3
DAY	FORAGE	MIN	MAX	DIR	B.S			HIVE WEIGHT		
1										
2										
3										
4										
5										
6										
7										
8										
9										
10										
11										
12										
13										
14										
15										
16										
17										
18										
19										
20										
21										
22										
23										
24										
25										
26										
27										
28										
29										
30										
31										

DEC17

1,FR

2,SA
SR:07:46, SS:03:53

3,SU ○
ADVENT SUNDAY

4,MO

5,TU

6,WE
ST NICHOLAS

7,TH
ST AMBROSE (PATRON SAINT OF BEEKEEPERS)

8,FR

9,SA
SR:07:54, SS:03:50

10,SU

11,MO

12,TU

13,WE

14,TH

15,FR

16,SA SR:08:01, SS:03:51	24,SU
17,SU	**25,MO** CHRISTMAS DAY
18,MO ●	**26,TU** BOXING DAY
19,TU	**27,WE** BANK HOLIDAY
20,WE	**28,TH**
21,TH WINTER SOLSTICE - FIRST DAY OF WINTER	**29,FR**
22,FR	30,SA SR:08:06, SS:03:59
23,SA SR:08:05, SS:03:51	31,SU NEW YEAR'S EVE

Hive/ Q NO.	Year Q Raised	Frames of Brood Autumn 2016	Combs Covered	Honey Stored- Sugar fed Kg	Combs Covered Spring 2017	Frames of Brood Spring 2017	Spring Feeding Kg	Queens Reared	Nuclei
1									
2									
3									
4									
5									
6									
7									
8									
9									
10									
11									
12									
13									
14									
15									
16									
17									
18									
19									
20									
21									
22									
23									
24									

HONEYBEE COLONIES

1									
2									
3									
4									
5									
6									
7									
8									
9									
10									
11									
12									
13									
14									
15									
16									
17									
18									
19									
20									
21									
22									
23									
24									

BEEEKEEPING RECORDS

Number	items	Est. Value £	P
	Stocks of Bees		
	Empty Hives		
	Combs - Deep - Shallow		
	Frames		
	Foundations		
	Honey Extractor		
	Honey Tanks		
	Other items		
	Honey Jars		
	Honey		

JANUARY 2018

S	M	T	W	T	F	S
	1	2	3	4	5	6
7	8	9	10	11	12	13
14	15	16	17	18	19	20
21	22	23	24	25	26	27
28	29	30	31			

FEBRUARY 2018

S	M	T	W	T	F	S
				1	2	3
4	5	6	7	8	9	10
11	12	13	14	15	16	17
18	19	20	21	22	23	24
25	26	27	28			

MARCH 2018

S	M	T	W	T	F	S
				1	2	3
4	5	6	7	8	9	10
11	12	13	14	15	16	17
18	19	20	21	22	23	24
25	26	27	28	29	30	31

APRIL 2018

S	M	T	W	T	F	S
1	2	3	4	5	6	7
8	9	10	11	12	13	14
15	16	17	18	19	20	21
22	23	24	25	26	27	28
29	30					

MAY 2018

S	M	T	W	T	F	S
		1	2	3	4	5
6	7	8	9	10	11	12
13	14	15	16	17	18	19
20	21	22	23	24	25	26
27	28	29	30	31		

JUNE 2018

S	M	T	W	T	F	S
					1	2
3	4	5	6	7	8	9
10	11	12	13	14	15	16
17	18	19	20	21	22	23
24	25	26	27	28	29	30

JULY 2018

S	M	T	W	T	F	S
1	2	3	4	5	6	7
8	9	10	11	12	13	14
15	16	17	18	19	20	21
22	23	24	25	26	27	28
29	30	31				

AUGUST 2018

S	M	T	W	T	F	S
			1	2	3	4
	6	7	8	9	10	11
	13	14	15	16	17	18
	20	21	22	23	24	25
	27	28	29	30	31	

SEPTEMBER 2018

S	M	T	W	T	F	S
						1
2	3	4	5	6	7	8
9	10	11	12	13	14	15
16	17	18	19	20	21	22
23	24	25	26	27	28	29
30						

OCTOBER 2018

S	M	T	W	T	F	S
	1	2	3	4	5	6
7	8	9	10	11	12	13
14	15	16	17	18	19	20
21	22	23	24	25	26	27
28	29	30	31			

NOVEMBER 2018

S	M	T	W	T	F	S
				1	2	3
4	5	6	7	8	9	10
11	12	13	14	15	16	17
18	19	20	21	22	23	24
25	26	27	28	29	30	

DECEMBER 2018

S	M	T	W	T	F	S
						1
2	3	4	5	6	7	8
9	10	11	12	13	14	15
16	17	18	19	20	21	22
23	24	25	26	27	28	29
30	31					

Every effort is made to keep entries up to date but the publishers cannot be held responsible for errors or omissions.
Associations and all other groups listed have been requested (August 2016) to supply updated entries.
Readers who are aware of inaccuracies are asked to send updates to jerry@northernbeebooks.co.uk

DIRECTORY, Associations and Services

DIRECTORY, ASSOCIATIONS AND SERVICES

Bee Educated (BEE-EDU)
e-Learning for Beekeepers

-†beekeeping virtual classroom.

Contact Us
Bee-Edu Administrator
Steven Turner
Email: st@zbee.com

Bee-Edu is a Moodle website set-up specifically for beekeeping tutors and their students.

What is Moodle?
Moodle is a Virtual Learning Environment (VLE) which makes it easy for tutors to provide online support for†their course. It provides†a central space on the web where students can access a set of tools and resources at all times.

What are the Benefits?
1. It is an easy way to communicate with students: The course news automatically emails messages to all students. Forums can also be used to answer commonly asked questions, to provide a space for informal peer to peer student discussion or even online tutorials.

2. It is at quick way to share documents: Moodle provides a place where you can easily create web pages with information about your course and provide links to word, PDF documents, slides, and other resources that your students will want to access.

3. It has easy access to relevant and useful online resources: There is so much information about beekeeping on the Internet which makes it difficult for students to find reliable and trusted resources. You can use your Moodle to provide links directly to these resources in an organised way.

4. Online assignment handling: Online assignment handling can save time and effort for everyone involved, whether itís just used for student submission with marking

done on paper or the whole process is moved online saving time, postage and paper.

5. Other advantages: Making resources available online can save time and money in photocopying. For example keep a central copy of documents†online so everyone can find the latest version of a course handbook etc. Provide handouts online so that students only print out what they really need. It is easy to experiment with new ideas and tools, it's a low risk way to incorporate new tools and ideas into your teaching. Tutors can manage their materials. If all your course information is on Moodle this is easy access this year on year.

6. Other features and tools: Course calendar: use this to flag important events to everyone on your course. Profiles and contact information helps students and tutors get to know each other from the start of the course. Deliver content: add slides and photographs. Video and audio: many tutors find it easy to record lectures as podcasts or even arrange for videos of lectures, posting these online and making it available to students is straight forward with Moodle.

7. Group tools for students. There are many tools that students can use for collaboration with each other such as forums, wiki and chat.

Beekeeping Course Tutors
Bee-Edu are providing a free virtual learning environment (VLE) for beekeeping tutors to complement their existing courses or build new online courses in beekeeping related topics.

Bee-Edu will support those individuals or associations that are keen to try this medium but help will be limited to testing and guidance with the technology rather than organising the teaching resources. There is no time scale and I will not be imposing any rules on tutors who wish do their own thing.

Students
It is†expected that most courses created on Bee-Edu would need enrolment keys and their students would be directed to the site by their tutors.

BEE DISEASES INSURANCE LTD

SECRETARY
Donald Robertson-Adams
Ffosyffin, Ffostrasol,
Llandysul, Ceredigion,
SA44 5JY
07532 336076
secretary@beediseases-insurance.co.uk

TREASURER AND SCHEME B MANAGER
Mrs Sharon Blake
Stratton Court,
South Petherton,
Somerset TA13 5LQ
01460 242124
treasurer@beediseases-insurance.co.uk

CLAIMS MANAGER
Bernard Diaper
57 Marfield Close,
Walmley,
Sutton Coldfield B76 1YD
07711456932
claims@beediseasesinsurance.co.uk

PRESIDENT
Martin Smith
137 Blaguegate Lane
Lathom, Sklemersdale
Lancs WN8 8TX
07831 695732
president@beediseases-insurance.co.uk

BDI is a small insurance company that specialises in compensating beekeepers in England and Wales, who have had their equipment destroyed by the Bee Inspector as a result of being infected by a notifiable disease. These currently are European Foul Brood (EFB) and American Foul Brood (AFB). BDI also has a fund to cover destruction as the result of being infested by a notifiable pest (Small Hive Beetle or Tropilaelaps), should they reach England or Wales.

The company is regulated as an insurance company by the Prudential Regulatory Authority and supervised by the Prudential Regulatory Authority and the Financial Conduct Authority.

BDI is owned by most BKAs who are their members. There are no full time employees or premises. BDI is run by a small group of officers on a day–to–day basis from their homes. In addition there is a board of directors who meet regularly.

BDI subscriptions are paid along with the local BKA subscription. This is compulsory if your BKA is a BDI member. You will also be asked to pay premiums for the number of additional colonies you expect to have during the year, above the basic free colonies although you can also top-up during a year.

Further details of current subscription and premium rates together with compensation rates are available on the BDI website.

Research Grants

BDI supports research into the causes of honey bee

diseases in a number of ways including the sponsorship of PhD studentships. Details of the current and past research activities that have been supported by BDI are available on the website.

Further details of the full range of activities carried out can be found on the BDI website www.beediseasesinsurance.co.uk, or contact via secretary@beediseasesinsurance.co.uk

Bee Farmers Assocation Limited

The BFA represents the professional beekeepers of the UK.
www.beefarmers.co.uk

CHAIRMAN
David Wainwright
Tropical Forest Products
Box 92
Aberystwyth
Dyfed
Wales
SY23 1AA
01970 832511
chair@beefarmers.co.uk

COMPANY SECRETARY
John Howat
8 Olivers Close
West Totton
Southampton
Hampshire
SO40 8FH
023 8090 7850
honsec@beefarmers.co.uk

FINANCE DIRECTOR
John Heard
36 The Green
Long Whatton
Loughborough
Leicestershire
LE12 5DB
01509 646767
financedirector@
beefarmers.co.uk

As the professional trade association for the sector, the Bee Farmers' Association represents around 450 bee farming businesses. Its members produce honey throughout Great Britain and supply products bulk, wholesale and retail. The association is the largest contract pollinator in the UK.

A significant number of members are employed as bee inspectors, responsible for identifying and dealing with notifiable disease.

There is one business meeting a year which follows the Annual General Meeting, held in the spring to coincide with one of the major trade events. There are also twice-yearly regional meetings, usually featuring guest speakers.

The Bee Farmers' Association works with the National Farmers' Union (NFU) and the Honey (Packers) Association.

The work of the Bee Farmers' Association

- To monitor and to keep members informed about developments in commercial beekeeping, bee science and UK and EU legislation.
- Liaison with farmers, growers, contractors, consumers and other organisations.
- Liaison and cooperation with UK beekeeping organisations.
- Liaison with UK Government departments dealing with beekeeping, medicines and allied matters.
- Contact with European beekeeping organisations and representation on the EU Honey Working Party (COPA/COGECA) in Brussels.
- Member of the EU Honey Task Force.
- Political lobbying through MPs and MEPs.
- Member of the Confederation of National Beekeeping Associations (CONBA).
- Associate member of the Honey Association.
- Associate member of the National Farmers' Union.

Benefits of membership

- Advice and support on all aspects of honey farming and commercial beekeeping.
- Networking opportunities with others in the sector.
- Insurance for products, third-party liability, and employer's liability.
- Association journal, featuring informative articles, case studies, news and updates on meetings with Defra, Fera, APHA, VMD and the EU, reports on current beekeeping issues and commercial developments worldwide.
- Free classified advertising and discounted display advertising in the association journal.
- Sources of equipment and sundries; product directory of specialist suppliers.
- Supplier discounts through BFA Sales. Discount vehicle purchase scheme.
- Regional meetings which provide networking and trading opportunities.
- UK Spring Conference and overseas visits; these include visits to bee farms and research establishments; lectures and discussions on bee-related matters; sight-seeing and social events.
- Eligibility for the Disease Accreditation Scheme for Honeybees (DASH).
- Free circulation of UK and foreign beekeeping journals.
- Crop and winter loss reports.
- Pollination contracts.

For Membership

Eligibility requires applicants to demonstrate they are developing or operating a bee farming business, or are employed in or working on behalf of a sector-related business or organisation.

Corporate Sponsorship

Please contact the General Secretary for details.

How to apply

Apply online at www.beefarmers.co.uk

GENERAL SECRETARY
Margaret Ginman MBA FRSA
Hendal House
Groombridge
Kent
TN3 9NT
01892 864499,
07795 153765
gensec@beefarmers.co.uk

Enquiries in relation to: European Union (EU) and government representation, strategic partnerships, corporate sponsorship, press and public relations, apprenticeship scheme.

MEMBERSHIP AND ADMINISTRATION
Alex Ellis
The Holding
Hanmer Road
Eglwys Cross
Whitchurch
Wrexham
Wales
SY13 2JP
01948 510726
admin@beefarmers.co.uk

Enquiries in relation to: membership, publications, services for members, insurance, events, Disease Accreditation Scheme for Honeybees (DASH).

POLLINATION SECRETARY
Alan Hart
61 Fakenham Road
Great Witchingham
Norwich
Norfolk
NR9 5AE
01603 308911
07867 523977
pollination@beefarmers.co.uk

BFA

Bee Farmer

Bee Farmer is the official journal of the Bee Farmers' Association and is the only United Kingdom (UK) journal focused solely on the interests of those involved in professional beekeeping. The emphasis of the publication is on practical issues relating to all aspects of commercial activity. UK, Europe and Rest of World subscription rates are available. Magazines are published in hardcopy format six times a year. Subscribe online at: www.beefarmers.co.uk

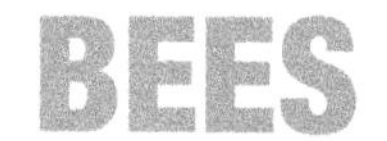

BEEKEEPING EDITORS' EXCHANGE SCHEME

BEES is a self-help grouping of local, county and country beekeeping association editors, which operates principally by exchanging journals through a central address. The scheme is supported by Northern Bee Books.

BEES was founded in 1984 and for many years has been an exchange of paper copy. However, the focus has now changed to an electronic exchange, using the server of one of the participating editors.

Now fully established as part of the British and Irish beekeeping scene, the scheme brings up to date information to beekeepers throughout the British Isles.

B.E.E.S
Helping Editors
Help Themselves

Sponsored by
NORTHERN BEE BOOKS

The aims are:

- to exchange ideas for content and production methods
- to aid others by experience
- to communicate matters editorial
- to share information on national beekeeping issues
- to help and reassure those new to the task
- to give a wider readership to the best writing in beekeeping journalism

If you are an editor or potential editor and would like to know more about how we operate write to Chris Jackson
22 Chapter Close, Oakwood, Derby, DE21 2BG
editors-owner@ebees.org.uk

BEES ABROAD

Relieving Poverty through Beekeeping

ADMINISTRATOR:
MRS VERONICA BROWN,
PO Box 2058,
Thornbury
Bristol
BS35 9AF.
0117 230 0231
info@beesabroad.org.uk

Bees Abroad is a UK-registered charity (No 1108464) established in 1999. Its principle aim is the relief of poverty in the developing world using beekeeping and associated skills as tools of individual, group and community empowerment for poverty alleviation and to provide sustainable income. Beekeeping is a valuable tool as it is socially and culturally acceptable for both genders across a wide age range. It can cost very little to set up a beekeeping operation, which will deliver benefits for income, education, health, environment and community. Beekeeping and its associated skills deliver access to gainful self-employment for poor and disadvantaged groups. This enables them to recover social status, improve social interactions, obtain income and acquire new skills to build the confidence to represent their own interests. Bees Abroad receives a high volume of direct appeals for assistance from groups all over the world. In practice, it achieves its aims through a volunteer network of supporters, committee members and project managers. Bees Abroad takes care to ensure that its projects are sustainable and not dependent on constant external input. This is done by supporting community group initiatives, setting up village-based field extension services, running training courses for beekeeping trainers and financing local trainers' wages. All Bees Abroad projects are designed to become self-financing after a defined time period. Its first two projects in Nepal and Cameroon now employ 42 beekeeper trainers and involve many more beekeepers. It currently has projects either running or seeking funding in Cameroon, Ghana, Kenya, Liberia, Malawi, Nigeria, Uganda, Yemen and Zambia.

Bees Abroad is run by volunteers who are all beekeepers. They currently undertake all activities, including fundraising, though an administrator is employed for one day a week. It also arranges beekeeping holidays to a variety of locations, including Morocco, Chile and Nepal.

Bees Abroad is delighted to have the support of its patrons: The Most Reverend Justin Welby, Archbishop of Canterbury, Professor Adam Hart (University of Gloucestershire), The Master of the Worshipful Company of Wax Chandlers, Martha Kearney (broadcaster and journalist), Michael Badger, MBE (past-president, BBKA), Brian Sherriff (BJ Sherriff International) and Eric Hiam (Maisemore Apiaries).

For more details of what we do and how you can help, contact Veronica Brown, the Administrator, Bees Abroad (info@beesabroad.org.uk). You can learn more about our work and make a regular or one-off donation through our website, www.beesabroad.org.uk

Bees for development

www.beesfordevelopment.org

YOU CAN HELP US BY:

- Encouraging your group or organisation to learn about and support our work
- Gifting a *Resource Box* for a training course in a school or project
- Subscribing to our quarterly beekeeping magazine, *Bees for Development Journal*.
- Sponsoring a *Journal* subscription to help a beekeeper working in a poor country
- Giving a donation
- Helping us to represent our organisation at events
- Offering your skills to work with us as a volunteer
- Joining one of our *Beekeepers' Safaris*
- Using our tamper proof seals or labels when you sell or gift your honey
- Buying from our *shop in Monmouth* or from our on-line store – proceeds from sales go to support our charitable development work.
- Attending one of our unique training Courses in UK on *Sustainable Beekeeping* or on *Strengthening Livelihoods by Means of Beekeeping*.

Bees for Development is the leading international charity organisation specialising in poverty alleviation through sustainable beekeeping.

We help some of the poorest communities in the world improve their well-being by providing the skills and knowledge that enable them to help themselves. We provide practical training in beekeeping, honey harvesting and the marketing of honey and beeswax products. Other issues we address in developing countries include the misuse of agrochemicals and poor land rights – these harm bees and hinder beekeepers trying to earn a living.

In 2016 our work in Africa encompassed projects in Cameroon, Eritrea, Ethiopia, Ghana and Uganda; previous years have seen similar work undertaken in Chechnya, Kyrgyzstan and India.

Bees for Development are as active in the UK as they are around the world. We are a key partner in the 'Bee Friendly Monmouthshire' campaign to retain wildflowers in hedgerows and verges for the benefits of bees and other pollinating insects. We also participate in local and national efforts to protect the environment, bees and their habitat and run a range of training courses in the Wye Valley area of south-east Wales.

Bees for Development Trust, charity no: 1078803

BRITISH BEEKEEPERS' ASSOCIATION

www.bbka.org.uk

COMMITTEES OF THE EXECUTIVE AND SECRETARIES

Education & Husbandry
The development of information from practical guidance notes, advisory leaflets, training materials while also undertaking it's own educational initiatives in support of improving the knowledge and skills of beekeepers at all levels. Education & Training liaise with the Examinations Board to develop training materials to support Association tutors with products such as the Course in a Case.

Examinations Board
The BBKA examination board provide a structured range of examinations fulfilling the needs of all beekeepers from Junior Certificate to Master Beekeeper. The board are responsible for all matters relating to the syllabus, content and assessment and operate independently of the BBKA board of Trustees. Where Associations have no Examinations Secretary the Association Secretary deals with examinations. To help future candidates it is suggested that Associations without an Examination Secretary appoint one. Associations are responsible for arranging a suitable room for the written examinations and recommending an invigilator.
Contact Val Francis, Exam Board Secretary
Email: val.frances@bbka.org.uk Tel 01226 286341

FINANCE
This team of Trustees reviews & agrees all budgets, handles all investment matters, finalising insurance policies and sets proposals relating to capitation.

OPERATIONS DIRECTOR & GENERAL SECRETARY
Vacant

BRITISH BEEKEEPERS ASSOCIATION
National Beekeeping Centre
Stoneleigh Park,
Kenilworth, Warwickshire
CV8 2LG
02476 696679
Fax: 02476 690682
Office hours 9.00am–5.00 pm
Monday - Friday (inclusive)

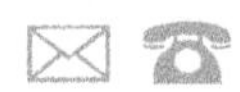

PRESIDENT
JOHN HENDRIE
john.hendrie@bbka.org.uk

CHAIRMAN
MARGARET MURDIN
margaret.murdin@bbka.org.uk

VICE CHAIRMAN
RUTH HOMER
ruth.homer@bbka.org.uk

TREASURER
CHRIS LAKE
treasurer@bbka.org.uk

Governance
Primary areas of responsibility are to ensure that we adhere to Charity Commission rules, that we operate within the constitution in addition to ensuring that our Trustees act in the best interests of the BBKA and it's members.

Operations & Membership Services
This team ensures that all Membership Services are administered effectively and on time and that the organisation operates efficiently. It also acts as a co-ordinator for all external fundraising.
Contact: jane.moseley@bbka.org.uk

Public Affairs
Whether it be government liaison, both UK & EU, or press activity this comes from the Public Affairs team. All enquiries should be made to BBKA Press Officer: diane.steele@bbka.org.uk

Technical & Environmental
All technical issues and their potential impact on bees and beekeeping are assessed and monitored within this team. All research projects are reviewed and recommendations made by Technical & Environmental group.

Insurance
Members of BBKA, Area Associations and officials are indemnified against claims for Public Liability to a limit of £10million, Product Liability to a limit of £10 million, Professional Indemnity to a limit of £2 million relating to their beekeeping activities. BBKA Association Officer and Trustee liability insurance also applies to a limit of £10 million. Each new claim carries an excess payable by the member.

An 'All Risks' policy is available to both individuals and Associations, to cover the loss or damage of property & equipment. Details are available via www.bbka.org.uk or the main office: 02476 696679

Publications

- BBKA News is issued monthly free to all members of the BBKA, featuring articles about bees, beekeeping and the other associated articles of interest. Editorial: editorial@bbkanew.org.uk Advertising: advertising@ bbkanews.org.uk
- BBKA Year Book is published each June and is for Association use and reference. It contains detailed information on the BBKA including useful reference tools such as a directory of Lecturers and Demonstrators.
- Members Handbook is published annually and sent to Association and Branch Chairman, Secretary, and Treasurer
- BBKA Introduction to Beekeeping

BBKA Website - www.bbka.org.uk

The BBKA Website contains technical information, is easy to navigate and supports both beekeepers and the general public. You can download publications, find help and advice in the discussion forums, purchase merchandise, learn about Bees, use the Bees4kids section, download BBKA exam application forms and the exam syllabus. Within the Members Only area, specific insurance downloads and other member only information is available. Associations beekeeping events are promoted.

A Swarm Collector database is included within the site enabling the general public with a direct link to a local swarm collector.

Events

Area and local associations attend and exhibit at various events within their local throughout the year while the BBKA supports selected national shows. Whether it be village fete or national exhibition these events continue to provide a vital service for the dissemination of knowledge.

TRUSTEES

Ken Basterfield
ken.basterfield@bbka.org.uk

Ian Homer
ian.homer@bbka.org.uk

Gareth Morgan
gareth.morgan@bbka.org.uk

David Teasdale
david.teasdale@bbka.org.uk

Howard Pool
howard.pool@bbka.org.uk

Simon Cavill
simon.cavill@bbka.org.uk

Margaret Wilson
Margaret.wilson@bbka.org.uk

Doug Brown
dougbrown@bbka.org.uk

BBKA Spring Convention
Held in April every year this is a firmly established major beekeeping event. Lectures and Workshops are staged over 3 days with a trade exhibition. Both Friday and Sunday are member only days which are ticketed. New in 2015 was the instigation of a Trade Day only ticket

Slide Library
The BBKA slide library has been digitised for ease of use and preservation. For a list of slides available and their format please go the BBKA Members Area at www.bbka.org.uk or contact the BBKA office.

Subscriptions & Membership Fees
Individual Membership of the BBKA is £38 per annum, for an Overseas Member the fee is £28.00. All other membership is via local associations.
Friends of the Honeybee Membership is also available via www.bbka.org.uk

Exam Board Footnote
Where Associations have no Examinations Secretary the Association Secretary deals with examinations. To help future candidates it is suggested that Associations without an Examination Secretary appoint one. Associations are responsible for arranging a suitable room for the written examinations and recommending an invigilator.

If you live in an area without a nominated Exam Secretary, you should contact Val Frances, Exam Board Secretary Email: val.frances@bbka.org.uk Tel 01226 286341

BBKA Enterprises
BBKA Enterprises Ltd is a private company, limited by guarantee with all profits from the trading activities being donated to the BBKA. Via the BBKA online shop a range of beekeeping, corporate and related items, specially selected books, gifts, travel items and educational materials are available.

Visit www.thepollenbasket.com, the official BBKA web shop, or call 02476 696679

SOME ASSOCIATION EXAM SECRETARIES

AVON
Mr Neil Seymour
The Old School House
Litton
Radstock
BA3 4PW
07921 256120
neil.seymour@gmail.com

BERKSHIRE
Mr John White
6, Horseshoe Road
Pangbourne
Reading
Berkshire
jkwhite70@hotmail.co.uk

BOURNEMOUTH & DISTRICT
Mrs Margaret Davies
57 Leybourne Avenue
Ensbury Park
Bournemouth
BH10 6HE
01202 526077
marg@jdavies.freeserve.co.uk

BUCKINGHAMSHIRE
Mrs Fiona Matheson
17 Shire Lane
Chorleywood
Hertfordshire
WD3 5NQ
01923 285637
education@buckscounty-beekeepers.co.uk

CAMBRIDGESHIRE
Mrs Eleanor Witter
177 Hills Road
Cambridge
CB2 8RN
01223 247228
eleanor.witter@tesco.net

CHESHIRE
Mr Graham Royle
7 Symondley Road
Sutton
Macclesfield
SK11 0HT
01260 252 042
g.royle@tiscali.co.uk

CHESTERFIELD
Mrs June Harvey
125 North Road
Clowne
Chesterfield
Derbyshire
S43 4PQ
01246 812115
harveyjex@aol.com

CORNWALL
Mrs Sue Malcolm
Figtree
333 New Road
Saltash
Cornwall
PL12 6HL
01752 845496
suzan@hmalcolm.freeserve.co.uk

CORNWALL WEST
Mrs Barbara Barnes
Clowance Barton Lodge
Praze-an-Beeble
Camborne
Cornwall
TR14 0PR
07901 977597
bab@barbara-barnes.com

CUMBRIA
Mr Peter Matthews
4 Annan Hill
Back of the Hill
Annan
Dumfries
DG12 6TN
01461 205525
silverhive@hotmail.com

DERBYSHIRE
Mrs Thelma Robinson
72 Church Street
Ockbrook Wood
Derby
Derbyshire
DE72 3SL
01332 662567
thelmaelizabethrobinson@gmail.com

DEVON
Mrs Lea Bayly
Blowiscombe Barton
Milton Combe
Devon
PL20 6HR
01822 855292
lea.jones2@btinternet.com

DORSET
Mr Terry Payne
Brookmans Farm Bungalow
Dunns Lane
Ewen Minster
Dorset
DT11 8NG
01747 811251
twpayne@btinternet.com

DOVER & DISTRICT
Mrs Jackie Thomas
Quarry House
Agester Lane
Denton Nr Canterbury
Kent
CT4 6NR
01227 831235
jackieaucott@gmail.com

DURHAM
Mr George Eames
11 Sharon Avenue
Kelloe
Durham
DH6 4NE
07970 926250
beeseames@btinternet.com

ESSEX
Mrs Pat Allen
8 Franks Cottages
St Mary's Lane
Upminster
Essex
RM14 3NU
01708 220897
pat.allen@btconnect.com

GLOUCESTERSHIRE
Mr Bernard Danvers
120a Ruspidge Road
Cinderford
GL14 3AG
01594 825063
berniedanvers@hotmail.co.uk

GWENT
Mrs Janet Bromley
Upper Ty Hir
Monmouth Road
Raglan
Gwent
NP15 2ET
01291 690331
bromleyjan@hotmail.com

HAMPSHIRE
Mrs Jean Frost
5 Pound Close
Upper Wield
Alresford
Hampshire
SO24 9SH
01420 561136
jeanterry@uwclub.net

HARROGATE & RIPON
Mrs Judith Hart
Kintail
Brearton
Harrogate
North Yorkshire
HG3 3BX
01423 865873
jm22.r27@virginmedia.com

BBKA

HEREFORDSHIRE
Mrs Louise Sheppard
The Seggin
Eyton
Leominster
Herefordshire
HR6 0BZ
01568 616692
louisesheppard2@hotmail.com

HERTFORDSHIRE
Mr David Canham
171, London Road
Hertford Heath
Hertfordshire
SG13 7PN
07990 530826
hertsbeeexams@gmail.com

HUNTINGDONSHIRE
Mrs Barbara Woodbine
16 Montagu Gardens
Kimbolton
Cambridgeshire
PE28 0JL
01480 861383
barbara@woodbine16.plus.com

ISLE OF MAN
Mrs Janet Thompson
Cott ny Greiney
Beach Rd
Port St Mary
Isle of Man
IM9 5NF
01624 835524
jthompson@manx.net

KENDAL & SOUTH WEST-MORLAND
Mr. Mick Gander
52 Buttermere Drive
Kendal
Cumbria
LA9 7PH
07515 797899
mickgander@live.com

KENT
Miss S Sharon Bassey
61, Nunhead Lane
London
SE15 3TR
sharonbassey@hotmail.com

NORFOLK WEST AND KINGS LYNN
Mrs Judith Heal
Burgh Parva Hall
Melton Constable
Norfolk
NR24 2PU
01263 862569
judyheal@dsl.pipex.com

LANCASHIRE & NORTH WEST
Mrs Barbarara Roderick
39, Hawksworth Drive
Formby
Liverpool
L73 7EY
01704 877855
beebarb@hotmail.co.uk

LINCOLNSHIRE
Mr Michael Seal
70a, Westfield Drive
North Greetwell
Lincolnshire
LN2 4RB
01522 754435
mikeseal91@sky.com

LONDON
Mr Howard Nichols
45 Selsden Road
West Norwood
London
SE27 0PQ
07809 156185
howard@wrightandco.biz

LUDLOW & DISTRICT
Mr Andy Vanderhook
The Old Forge
Baveney Wood
Cleobury Mortimer
Kidderminster
DY14 8JD
01584 890830
andy.vanderhook@care-4free.net

MANCHESTER & DISTRICT
Mrs Joy Jackson
96a, New Lane
Eccles
Lancashire
M30 7JE
07872 512266
joy.jackson4@hotmail.co.uk

MEDWAY
Mr Terry Clare
89 Chalky Bank Road
Rainham
Gillingham
Kent
ME8 7NP
01634 233748
terryeclare@tinyworld.co.uk

MIDDLESEX
Mrs Jo Telfer
Midwood House
Elm Park Road
Pinner
Middx
HA5 3LH
020 8868 3494
jvtelfer@hotmail.com

MOLE APIARY CLUB
Mr Denis Cutler
70 Hurst Rd
East Molesey
Surrey
KT8 9AG
0208 224 9283
densicutler@ntlworld.com

NEWCASTLE & DISTRICT
Mrs Valerie Hawley
Tindall House
Killingworth Village
Newcastle upon Tyne
Tyne and Wear
NE12 6BL
01912 683949
val.hawley@btinternet.com

NORFOLK
Mrs Carolyne Liston
Ivy Cottage
Dumbs Lane
Hainford
Norwich
NR10 3BH
01603 893330
cliston@ukf.net

NORTHAMPTONSHIRE
Mr Mike Hall
3, Thorpeville
Moulton
Northamptonshire
NN3 7TS
halle3m@gmail.com

NORTHUMBERLAND
Mr Ian Robson
2 Breamish Gardens
Powburn
Alnwick
Northumberland
NE66 4HQ
07833 317399
Ian@kw-porvis.co.uk

NOTTINGHAMSHIRE
Ms Janet Bates
11, Rowan Avenue
Ravenshead
Nottingham
NG15 9GA
01623 794687
janet.bates@ntlworld.com

OXFORDSHIRE
Mr Peter Chaunt
9 Robins Close
Barford St Michael
Banbury
Oxford
OX15 0RP
01869 338625
chaunt@talktalk.com

PETERBOROUGH & DISTRICT
Mr George Newton
65 Queen Street
Yaxley
Peterborough
Cambs
PE7 3JE
01733 243349

SHROPSHIRE
Mrs Liz Williams
35 Ridgebourne Road
Shrewsbury
Shropshire
SY3 9AB
e.williams800@btinternet.com

SHROPSHIRE NORTH
Mrs Joyce Nisbet
22 Ffordd Ystrad
Coed y Glyn
Wrexham
Shropshire
LL13 7QQ
01978 363168
joycerussell1@hotmail.co.uk

SOMERSET
Mrs B Bridget Knutson
6 Wideatts Road
Cheddar
Somerset
BS27 3AP
01934 742187
bridget_knutson@yahoo.co.uk

STAFFORDSHIRE NORTH
Ms A Angela Fearon
The Crofters
The Green
Stocton Brook
Staffordshire
ST9 9PD
07764 605663
angelafearon@googlemail.com

STAFFORDSHIRE SOUTH
Mr Julian Malein
Woodview
School Lane
Admaston nr Rougley
Staffordshire
WS15 3NH
01889 500486
jmalein@yahoo.co.uk

STRATFORD ON AVON
Mr Terry Hitchman
Church View
Pillerton Hersey
Warwickshire
CV35 0QJ
01789 740136
terryhitchman@phonecoop.coop

SUFFOLK
Mr Adrian Howard
Rondebosch
Lodge Road
Hollesley
Woodbridge
IP12 3RR
01394 4111561
a.howard106@btinternet

SURREY
Mrs Celia Perry
White Gables
68 Broadhurst
Ashstead
KT21 1QF
0790 3991120.
beeexams@hotmail.co.uk

SUSSEX
Mrs Liz Twyford
Westcott
Udimore Road
Broad Oak
Rye
TN31 6DG
01424882361
secretary@sussexbee.org.uk

SUSSEX WEST
Mr Roger Brooks
23, Lionel Avenue
Bognor Regis
West Sussex
PO22 8LG
01243 584531
rhandm@hotmail.com

THANET
Mrs Rowena Pearce
Summerfield Cottage
Summerfield
Woodnesborough
Sandwich
CT13 0EW
01304 614789
pearcesinsummerfield@tiscali.co.uk

TWICKENHAM & THAMES VALLEY
Mr Chris Deaves
12 Chatsworth Crescent
Hounslow
TW3 2PB
02085 682869
c_deaves@compuserve.com

WARWICKSHIRE
Mr Bob Gilbert
66 Sharp Street
Amington
Tamworth
Staffordshire
B77 3HZ
01827 65749
bee1bob1@aol.com

WILTSHIRE
Mrs Sally Wadsworth
57 St Edith's Marsh
Bromham
Chippenham
Wiltshire
SN15 2DF
01380 859052
sally.wadsworth@btinternet.com

WORCESTERSHIRE
Mr Martin Cracknell
Honeylands
Abberton Road
Bishampton
Worcestershire
WR10 2LU
01386 462385
martyn@cracknellz.freeserve.co.uk

WYE VALLEY
Mrs Susan Quigley
Newhouse Farm
Michaelchurch Escley
Hereford
Herefordshire
HR2 0PT
01981 510183
quigley.susan@hotmail.co.uk

YORKSHIRE
Mrs Yvonne Kilvington
Membership
144 93 603 961
10 Banks Avenue
Golcar
Huddersfield
West Yorkshire
HD7 4LZ
01484 643314
ykilvington@btopenworld.com

SEDBERGH
Mr John Rogers
Holly Bank
Ingleton
Carnforth
LA6 3DR
01524 241364
johnrogers@btinternet.com

VALE & DOWNLAND
Mrs Lilian Valentine
6 Grove Road
Wantage
Oxfordshire
OX12 7BU
01235 767524
jvalentine515@btinternet.com

NEWBURY
Mr Michael White
17 Donnington Square
Newbury
RG14 1PJ
0163 544945
mpwhite@freegratis.net

CLEVELAND
Mr Tom Rettig
Hillcrest Village
Middleton-on-Leven
Yarm
TS15 0JX
01642 596158
t.rettig@btinternet.com

RUTLAND
Mr Will Rigby
21 Mill Street
Melton Mowbray
Leicestershire
LE13 1AY
01664 852742
will@gliderman.fsnet.co.uk

ISLE OF WIGHT
Mrs Liz Van Wyk
3, Buckingham Road
Ryde
Isle of Wight
PO33 2DP
01983 565839
elizabethvanwyk@aol.com

JERSEY
Mrs Judy Collins
2 Demerara Cottages
Le Mont Sohier
St Brelade
Jersey
JE3 8EA
07797 790420
judybees@collinsje.net

INSTITUTE OF NI BEEKEEPERS
Mr Tom Canning
151 Portadown Road
Armagh
BT61 9HL
07867 878474
tjcanning@btinternet.com

LANCASTER
Mr Peter Stephens
49, Redhills Road
Arnside
Cumbria
01524 761445
peterstephens7@btinternet.com

ASSOCIATION SECRETARIES

10	Ms	R	Taylor	43, High Street, Chew Magna	BRISTOL
11	Mr	M	Moore	19, Armour Hill, Tilehurst	READING
12	Mr	P	Darley	3 Dorset House, 42, The Avenue	POOLE
13	Mrs	S	Carter	74 Whitelands Avenue, Chorleywood	
14	Miss	S	Fenwick	27 Pratt Street, Soham	ELY
15	Mrs	E	Camm	Magpie Manor, Wistaston Green Road	CREWE
16	Mr	R A	Bagnall	21 Ramper Avenue, Clowne, Chesterfield	DERBYSHIRE
17	Mrs	A	Ramsden	1 Wellington Place, Old Carnon Hill, Carnon Downs	TRURO
18	Ms	K	Bowyer	The Nook, Lower Carnkie	REDRUTH
19	Mr	S	Barnes	8 Albermarle Street, Cockermouth	CUMBRIA
20	Mr	M J	Cross	Harlestone, Beggarswell Wood, Ambergate	DERBYSHIRE
21	Mr	B	Neal	Badgers Barn, Withacott, Langtree	TORRINGTON
22	Mrs	L	Rescorla	5, Cowleaze, Martinstown	DORCHESTER
23	Mrs	M E	Harrowell	4 Harton Cottages, Chapel Lane, Ashley	DOVER
24	Mrs	L	Ramsey	Ashes House Farm, Wolsingham	BISHOP AUCKLAN
25	Mr	M	Webb	19 Ingrebourne Gardens, Upminster	ESSEX
26	Mrs	R	Savage	Oak House, Windrush Gardens	LYDNEY
27	Mrs	G	Williams	Green Court, South Row, Redwick, Magor	CALDICOT
28	Ms	Z	Semmens	27 Oak Tree Drive, Hook	HAMPSHIRE
29	Ms	J	Paley	20, Kent Road	HARROGATE
30	Mr	C	Stowell	Clayfoot Farm, Linley Green Road, Whitbourne	WORCESTER
31	Mr	J	Palombo	Field End, Woodside Green, Great Hallingbury BISHOP'S STORTFORD	
32	Mrs	M	Watkin	2, Rusts Lane, Alconbury	HUNTINGDON
33	Mrs	P	Shimmin	66, Ormly Road, Ramsey	ISLE OF MAN
34	Mrs	N	Mumberson	91, Victoria Avenue	SHANKLIN
35	Mrs	J	Bayne	Helm End Farm, Barrows Green, Stainton	KENDAL

No.	Title	Initials	Surname	Address	Town
36	Mrs	J	Spon-Smith	77 Bushey Way	BECKENHAM
37	Mr	A	Davies	West Cottage, Swanton Morley Road, Worthing	DEREHAM
38	Mr	M	Smith	137 Blaguegate Lane, Skelmersdale	LANCASHIRE
39	Ms	P	Merriman	19, Greenacre Court	LANCASTER
40		S	Raines	Grange Cottage, 21 Humberstone Avenue	GRIMSBY
41	Ms	E	Nye	6 Mayfield House, Rushcroft Road	LONDON
42	Mr	A	Jenyon	Oakleigh, Oldwood Road	TENBURY WELLS
43	Mrs	J	Jackson	96a, New Lane, Eccles	MANCHESTER
44	Mrs	M	Hunter	18, Slades Gardens	ENFIELD
45	Mrs	R	Conway	2a, Tippings Lane, Barrowden	OAKHAM
46	Mr	V	Cassidy	10, Bankside Close, Ryhope	SUNDERLAND
47	Mrs	J	Harrison	Woodgate Barn, Frogs Hall Lane, Swanton Morley	DEREHAM
48	Mrs	R	Stewart	17, Leys Avenue, Rothwell	KETTERING
49	Mr	B	Hopkinson	11 Watershaugh Road, Warkworth	MORPETH
50	Mr	MS	Jordan	29 Crow Park Avenue, Sutton-on-Trent,	NOTTINGHAM
51	Miss	H C	Raine	37, St Marys Road, Adderbury	BANBURY
52	Mr	P G	Newton	65 Queen Street, Yaxley, Peterborough	CAMBS
53	Mrs	C	Currier	Churchleigh, Adderley, Market Drayton	SHROPSHIRE
54	Mr	N	Hine	Chapel House, Whixall	WHITCHURCH
55	Dr	R	Bache	11 Rectory Mews, Hatch Beauchamp	TAUNTON
56	Mr	S	Boulton	Middle Banks, Malthouse Road, Alton	STOKE-ON-TRENT
57	Mrs	L	Lacey	21, Fisherwick Road	LICHFIELD
58	Mr	MJ	Osborne	Oak Lodge, Kings Lane, Snitterfield STRATFORD-UPON-AVON	
59	Mr	I J	McQueen	643 Foxhall Road	IPSWICH
60	Mrs	S	Rickwood	19 Kenwood Drive, Walton-on-Thames	SURREY
61	Mrs	E	Twyford	Westcott, Udimore Road, Broad Oak	RYE
62	Mr	G	Elliott	Robins Croft, Chalk Road, Ifold Loxwood	BILLINGSHURST
63	Mrs	R	Pearce	Summerfield Cottage, Summerfield, Woodnesborough	SANDWICH
64	Ms	S	Crofton	10 Wellesley Avenue, London	
65	Mrs	G	Rose	40, Beaudesert Road	BIRMINGHAM
66	Mr	B	Wilson	71e, School Lane, Shaw	MELKSHAM
67	Mrs	L	Chapman	White House, Warbage Lane, Dodford	BROMSGROVE
68	Mrs	S	Quigley	Newhouse Farm, Michaelchurch Escley	HEREFORD
69	Mrs	C	Thomson	105, Cidercourt Road	CRUMLIN
70	Mr	R	Chappel	4, The Green, Brafferton	DARLINGTON
71	Mrs	SJ	Stunell	7, Marshland View, Lower Stoke	ROCHESTER
72	Mrs	J	Doyle	1, Cherry Orchard, Great Shefford	HUNGERFORD
73	Mrs	J	Greenhalgh	8, Park View, Garston Lane	WANTAGE

74	**Mrs**	**L**	**Pauley**	Croft House, Newby, Clapham	LANCASTER
75	**Mr**	**I**	**Makinson**	36 Montagus Harrier	GUISBOROUGH
76	**Ms**	**P**	**Kilduff**	Rowan House, 4, Le Clos St Sampson La Route Des Quennevais, St Brelade	JERSEY
77	**Mr**	**P**	**Lythgoe**	16 , Stockburn Drive, Failsworth	MANCHESTER
95	**Ms**	**L**	**Pearce**	27, The Forstal, Hadlow, tonbridge	OXFORDSHIRE

Where Associations have no Examinations Secretary the Association Secretary deals with examinations. To help future candidates it is suggested that Associations without an Examination Secretary appoint one. Associations are responsible for arranging a suitable room for the written examinations and recommending an invigilator.

If you live in an area without a nominated Exam Secretary, you should contact Mrs Val Frances,
39 Beevor Lane, Gawber, Barnsley, S75 2RP
Tel 01226 286341. e-mail, toval.francis@bbka.org.uk

HOLDERS OF THE BBKA SENIOR JUDGES CERTIFICATE

Mr. Terry Ashley
Meadow Cottage,
11, Elton Lane, Winterley,
Sandbach, Cheshire. CW11 4TN
01270760757

Mr. Dennis Atkinson
4, Fell View, Garstang,
Lancashire. PR3 1WQ
01995602058

Mr. Michael Badger MBE
'Kara', 14, Thorn Lane,
Roundhay, Leeds,
West Yorkshire. LS8 1NN
0113294 5879
buzz.buzz@ntlworld.com

Mrs. Hazel Blackburn
28 Chazey Road, Caversham, Reading,
Berkshire. RG4 7DS
01189475451

Mr Alan Brown
9 The Woodlands, Carleton, Pontefract
Yorkshire. WF8 2RN
01977 796193
alanhoneybees4u@talktalk.net

Mrs. Vivienne Brown
7, Links Way, Flackwell Heath,
High Wycombe,
Buckinghamshire. HP10 9LZ
01628521502

Mr. Martin Buckle
The Little House,
Newton Blossomville,
Bedford, Beds. MK43 8AN
01234881262

Rev'd Francis Capener
1, Baldric Road,
Folkestone, Kent. CT20 2NR
01303254579

Mr. Gerald Collins, NDB
72, Tatenhill Gardens,
Doncaster, Yorkshire. DN4 6TL
01302539873
gerry@collins72.plus.com

Miss Margery Cooper
10, Gaskells End,
Tokers Green, Reading,
Berkshire. RG4 9EW

Mrs. Moyra Davidson
Hazlefield House, Auchencairn,
Castle Douglas DG7 1RF
01556 640597

Mrs. Margaret Davies
57, Leybourne Avenue,
Ensbury Park, Bournemouth,
Dorset. BH10 6ES
01202526077
margaretdavies773@btinternet.com

Mr. Bernard Diaper
57, Marfield Close,
Walmley, Sutton Coldfield,
West Midlands. B76 1YD
07711 456932
b.diaper@tiscali.co.uk

Ms. Fiona Dickson
Didlington Manor, Didlington, Thetford,
Norfolk. IP26 5AT
01842878673

Mr. Mike Duffin
Upper Hurst, Salisbury Road,
Blashford, Ringwood,
Hampshire. BH24 3PB
01425474552

Mr. Leo Fielding
Linley, Station Road,
Lichfield, Staffordshire.
WS13 6HZ
01543264427

Mr. Ivor Flatman
15, Waterton Close,
Walton, Wakefield,
West Yorkshire. WF2 6JT
01924257089
ivor.flatman@homecall.co.uk

Mr. Stephen Guest
Bridge House,
Hindheath Road,
Wheelock, Sandbach,
Cheshire. CW11 9LY
01270762226

Mr John Goodwin
Foleshill, Brereton Heath
Congleton , Cheshire. CW12 4SY
01477535032
john.goodwin@virgin.net

Mrs. Mary Hill
Whittington, Selling Road,
Old Wives Lees,
Canterbury, Kent. CT4 8BH
01227730477
mary.hill43@btinternet.com

Mr. Michael MacGiolla
Glengarra Wood,
Burncourt, Cahir,
Co. Tipperary, EIRE.
0035352672053

Mr. Peter Matthews
4, Annanhill, Back of the Hill,
Annan, Dumfries & Galloway,
Scotland. DG12 6TN
01461 205525

Mr. Gerald Moxon
9, Savery Street,
Southcoates Lane,
Kingston upon Hull,
Yorkshire. HU8 8DG
01482782052

Mr. Jim Orton
Occupation Road,
Sibson, Nuneaton,
Warwickshire. CV13 6LD
01827880471

Mrs. Suzette Perkins
Tengore House, Tengore lane
Langport, Somerset
TA10 9JL
01458 250095

Wg Cmdr Tom
Salter MBE C.Eng RAF
Splash Hollow, Five Bells Lane
Nether Wallop, Stockbridge
S020 8EN
01264 781382
tomasalter@hotmail.co.uk

Mr David Shannon
April Court, High Street, Wroot,
Doncaster, South Yorkshire DN9 2BT
01302 772837
daveshannon.aca@me.com

Mr. Chris Symes
189, Marlow Bottom Road,
Marlow, Buckinghamshire.
SL7 3PL
01628485212

Mr. Redmond Williams
Tincurry, Cahir,
Co. Tipperary, EIRE.

Mr Alan Woodward
55, Smillie Road
Rossington, Doncaster
South Yorkshire, DN11 0AW
01302868169

Mr. Michael Young MBE
'Mileaway', Carnreagh,
Hillsborough, Northern Ireland.
BT26 6LJ
02892 689724
myoungjudge@yahoo.co.uk

Mr. GeorgeVickery*
'Ponderosa', Verwood Road,
Three Legged Cross,
Wimborne, Dorset. BH21 6RN
01202825774

* These judges are no longer active.

BEE IMPROVEMENT & BEE BREEDERS' ASSOCIATION (BIBBA)

www.bibba.com
Membership Secretary: Nick Mawby
membership@bibba.com

About BIBBA

BIBBA is an organisation of enthusiasts and supporters of native and near native honey bees. We encourage beekeepers to improve their bees by using queens that are raised by simple methods the bees often present us with, or more advanced methods if larger numbers of queens are required.

BIBBA was formed in 1964 and has continually helped and encouraged beekeepers to keep docile and productive bees that are calm on the comb, healthy and suit the environment they are in. This is done by staging events and providing a number of relevant publications that are available from the BIBBA website and at shows where BIBBA has a presence.

Events

Events can be staged throughout the year by working closely with local beekeeping associations (BKAs) to provide relevant tuition to beekeepers. These include "Bee Improvement For All" (BIFA) days that are held inside between October and March and one and two day bee improvement courses held outdoors during the summer.

THE C.B. DENNIS BRITISH BEEKEEPERS' RESEARCH TRUST

REGISTERED CHARITY NO. 328685

This independent Charitable Trust has now been in existence for 25 years. During that time almost £500,000 has been awarded in grants for research in support of bee science. From small beginnings the Trust has now become a significant funding agency for bee research in this country awarding grants to universities and institutions on the basis of scientific merit and supporting young bee scientists by providing funding for studentships and training. Since its inception the Trust has funded work on a wide range of topics related to both honey bees and other bees, always aiming to support excellent science of benefit to bees and beekeepers in this country.

Awards

The Trust is administered by a group of seven Trustees all of whom are, or have been career scientists. They therefore have first-hand knowledge of both writing and evaluating research proposals and several have extensive practical experience of working with bees in a professional or hobbyist capacity. This expertise ensures that work funded by the Trust is properly evaluated and provides the greatest possible advantage for bees. Meetings are held twice a year in April and October to evaluate submitted research applications.

Donations

The Trust is pleased to acknowledge the loyal support it already receives from several local beekeeping associations and many individuals. All donations, however small, will be added to the invested capital and bee research in Britain will benefit from the income in perpetuity. Full details of the activities of the Trust, outputs of the research funded and grant application forms can be obtained from www.cbdennistrust.org.uk

HON. SECRETARY, MS B.V.BALL
104 Lower Luton Road
Wheathampstead
St. Albans
Herts AL4 8HH

TREASURER, MS J HUMBY

ALL DONATIONS AND CORRESPONDENCE SHOULD BE SENT TO THE SECRETARY.
Email:
cbdennisbeetrust@gmail.com

THE CENTRAL ASSOCIATION OF BEEKEEPERS

www.cabk.org.uk

SECRETARY, Pat Allen
8 Frank's Cottages
St Mary's Lane
Upminster, RM14 3NU
pat.allen@btconnect.com

PRESIDENT, Prof. R.S. Pickard
pickard.r@btopenworld.com

TREASURER, Harold Cloutt
Corriemulzie
Netherfield
Battle
Sussex TN33 9PY
bees@cloutthr,plus.com

PROGRAMME SECRETARY
Pam Hunter
Burnthouse
Burnthouse Lane
Cowfold, Horsham
RH13 8DH
pamhunter@burnthouse.org.uk

SALES AND DISTRIBUTION,
Bill Fisher
The Old Farmhouse,
Farm Road, Chorleywood,
Hertfordshire
WD3 5QB
theoldfish@hotmail.com
07973 626464

The Central Association of Beekeepers in its present form dates from the time of the reorganisation of the British Beekeepers' Association in 1945. The BBKA was originally made up of private members only. However as County Associations were formed they applied for affiliation and were later permitted to send delegates to meetings of the Central Association, as the private members were then known. This arrangement became unsatisfactory as the voting power of the Central Association greatly outnumbered that of the County Associations and so in 1945 a new Constitution was drawn up whereby the Council comprised Delegates from the Counties and Specialist Member Associations. The private members then formed themselves into a Specialist Member Association with the designation 'The Central Association of the British Beekeepers' Association'; this was later shortened to its present style.

The Association was able to devote itself to its own particular aims, to promote interest in current thought and findings about beekeeping and aspects of entomology related to honey-bees and other social insects. Lectures given by scientists and other specialists are arranged, printed and circulated to members, as has been done since 1879.

A Spring Meeting with three lectures plus Annual General Meeting is held in London, and an Autumn Weekend Conference in the Midlands. In addition a lecture is given at the Social Evening held during the National Honey Show. Subscriptions are £15 per annum for an individual, £18 for dual membership, £20 for corporate membership.

THE EVA CRANE TRUST

www.evacranetrust.org
mail@evacranetrust.org
@evacranetrust

Eva Crane Trust

Trust Chairman
Richard Jones

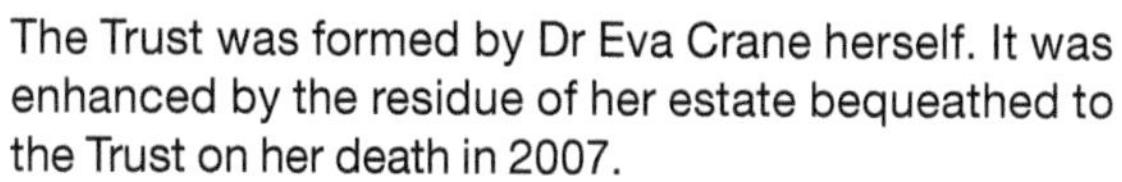

The Trust was formed by Dr Eva Crane herself. It was enhanced by the residue of her estate bequeathed to the Trust on her death in 2007.

The aim of the Trust is to continue Dr Crane's work in the way she would have liked it to evolve. This includes advancing the understanding of bees and beekeeping by the collection, collation and dissemination of science and research worldwide, as well as recording and propagating a further understanding of beekeeping practices through historical and contemporary discoveries.

The Trust, as well as being Dr Crane's way of ensuring her work continues, is a memorial whereby it may be possible to help fund others who can build on the foundations of sound academic research laid down in her many publications. Grants may be made to individuals and organizations that might otherwise find funding difficult in this specialized field. Applications will be considered from anywhere in the world but must be made in writing in the English language, preferably using the form on the website.

A comprehensive webssite continues to expand with new information on a regular basis. It is intended to be a research portal particularly for those interested in the history and development of bee science and beekeeping.

http://www.EvaCraneTrust.org

Similar information can be obtained by writing to:
The Eva Crane Trust, c/o Withy King Solicitors,
5-6 Northumberland Buildings, Bath, BA1 2JE, UK
Email: mail@evacranetrust.org

CONBA

✉ ☎

CONBA-UK & Ireland

COUNCIL OF NATIONAL BEEKEEPING ASSOCIATIONS IN THE UNITED KINGDOM and IRELAND

Incorporating the beekeeping organisations of: England, Channel Islands Isle of Man, Wales, Scotland, Ulster, Ireland and The Bee Farmers Association

SECRETARY
Phil McAnespie
12 Monument Road
Ayr KA7 2RL
01292 885660
philmcanespie@btinternet.com

CHAIRMAN,
Margaret Ginman
Hendal House
Groombridge
Kent
TN3 9NT

VICE CHAIR
Bron Wright,
20 Lennox Row
Edinburgh
EH5 3JW

TREASUER
Phil McAnespie
12 Monument Road
Ayr KA7 2RL
01292 885660
philmcanespie@btinternet.com

CONBA was established in 1978 to promote the aims and objectives of the national beekeeping associations of England, Scotland, Ulster, Wales and Ireland, and the Bee Farmers Association. Its purpose is to represent the interests of beekeepers' with local, national and international authorities. A representative delegate from each of the member country associations occupies the chair for a period of two years, on a rotational basis.

The council meets twice per year, normally at the Spring Convention and at the National Honey Show in London. Council business consists of any matters of common interest to all its members. CONBA provides representation of its membership at the European Union (EU) through two specific committees, COPA and COGECA (COPA – Comite des Organisations Professionelles Agricoles de la CEE); (COGECA Comite de la Cooperation Agricole de la CEE); and the Honey Working Party (HWP).

The Honey Working Party meetings are held at Brussels. This committee liases with the European Commission in relation to apicultural matters concerning the member states of the European Union (EU). These matters are subsequently presented to the European Parliament for its consideration, implementation or revision or rejection. The subsequent approval of such matters results in establishing legislation, government support and possible EC funding relating to the practice of apicultural production in the UK through its membership of the EU.

COUNCILLORS REPRESENTING THE MEMBER ASSOCIATIONS

BFA: Margaret Ginman (Chair)
BBKA: Ian Homer
SBA: Bron Wright, Phil McAnespie
UBKA: Susie Hill, John Hill
FIBKA: Michael Gleeson,
WBKA: Jenny Shaw, John Bowles

DEVON APICULTURAL RESEARCH GROUP

CHAIRMAN, Richard Ball
Stoneyford Farmhouse
Colaton Raleigh
Sidmouth, Devon EX10 0HZ
01 395 567 356

HON SECRETARY, Vacant
Contact Chairman

TREASURER, Bob Ogden
Pennymoor Cottage
Pennymoor, Tiverton
Deven EX16 8LJ
01363 866687

D A R G is an independent group of experienced enthusiastic beekeepers whose primary aim is to collect and analyse data on matters of topical interest which may assist their apicultural education and promote the advancement of beekeeping. At their regular meetings, DARG members discuss various topics in open forum, during which they exchange ideas and information from their personal beekeeping knowledge and experience. They also undertake suitable research projects which further the Group's aims.

TOPICS CURRENTLY BEING UNDERTAKEN

- Use of management (mechanical) methods including shook colonies for varroa control.
- Brood cell size in natural comb.
- Pollen diversity in Devon and neighbouring counties.
- A survey of useful bee plants, shrubs and trees in the South West.
- Drone movement between colonies.

In conjunction with Devon BKA

- Survey of Nosema in the County of Devon.
- Survey of drone laying queens in the County of Devon.

PUBLICATIONS AVAILABLE

- **The Beeway Code.** A common sense guide for beginners to help avoid problems with neighbours and produce a safe and peaceful apiary. Available from Northern Bee Books
- **Seasonal Management**. Currently under review.
- **Queen Rearing.** Currently under review.
- **The selection of Apiary sites** Currently under review.

THE FEDERATION OF IRISH BEEKEEPERS' ASSOCIATIONS

http://www.irishbeekeeping.ie

Secretary: Mr Tom Shaw,
201 Ard Easmuinn, Dundalk, Co. Louth.
Email: secretary@irishbeekeeping.ie
Mobile: 086 236 1286 Home: 042 93 39 619

ANNUAL SUMMER COURSE
The dates for the 71st Summer Course are Sunday 30th of July to Friday 4th of August 2017 at the Franciscan College, Gormanston, Co Meath with guest speaker Dr Ralph Büchler from the Bee Institute Kirchhain, Kassel, Germany.

Summer Course Convenor: Mr Michael G Gleeson. Ballinakill, Enfield, Co Meath.
Tel No 046-9541433/087-6879584, email mgglee@eircom.net; or visit http://www.irishbeekeeping.ie/gormanston/gormprog2016.html

PUBLICATIONS:
Having Healthy Honeybees - Published by F.I.B.K.A. Editor John McMullan, Ph.D.
The aim of this book is to help beekeepers establish healthy honeybee colonies, assess their condition and take appropriate action. Diseases are dealt with in a concise format to improve readability and are referenced to the latest peer-reviewed research. The book emphasises the importance of proper set-up involving an integrated approach to health management, in effect a preventative system that comes at little extra cost to the beekeeper
Cost €15 + P & P of €2 each
Bulk buying available to Associations In packs of 10 or 20 books, available at €12 each + P & P of €10 for packs of 10 or 20.
The recommended price is €15 per copy.
It is highly recommended for those doing the various FIBKA Examinations.
Available from the Hon Secretary

OFFICERS:

President: Mr Gerry Ryan,
Deerpark, Dundrum,
Co Tipperary
Tel No 062-71274 &
087-1300751
E-mail president@
irishbeekeeping.ie

Vice President: Stuart Hayes,
54 Glenvara Park,
Knocklyon, Dublin 16
Tel: 085 160 2613
E-mail vice.president@
irishbeekeeping.ie

Treasurer: Ms Maria Tobin,
Curragh, Donoughmore,
Co. Cork.
Tel: 086-0888780
E-mail: treasurer@
irishbeekeeping.ie

Editor: Ms Mary Montaut,
4 Mount Pleasant Villas,
Bray, Co. Wicklow.
Tel: 01-2860497
E-mail:
editor.@irishbeekeeping.
ie

Manager:
Mr Dermot O Flaherty,
Rosbeg, Westport,
Co. Mayo.
Tel: 098-26585 and
087-2464045
E-mail: manager.
beachaire@
irishbeekeeping.ie

FIBKA

Education Officer:
Michael Maunsell
Clonegannagh, Dunkerrin,
Birr, Co. Offaly.
Tel: 050 545 340 &
087 413 1622
E-mail: education.officer@
irishbeekeeping.ie

Summer Course Convenor:
Mr Michael G Gleeson,
Ballinakill, Enfield,
Co. Meath.
Tel: 046-9541433 and
087-6879584,
E-mail: mgglee@eircom.net

P R O: Paul O'Brien,
Fortlodge,Treanlaur-Maree,
Oranmore, Co. Galway.
Tel: 087 987 1800
E-mail: pro@
irishbeekeeping.ie

Bee Health officer:
Eleanor Attridge
Ballinaskeha, Leamlara,
Co. Cork.
Tel: 087-6879584
E-mail: beehealth@
irishbeekeeping.ie

Life Vice-Presidents:
Mr P. O'Reilly,
11 Our Lady's Place,
Naas, Co Kildare
Tel: 045-897568
E-mail: jackieor@indigo.ie

Mr Ml Woulfe,
Railway House, Middleton,
Co Cork
Tel: 021-4631011
E-mail: glenanorehoney@
eircom.net

President ex-officio
Mr. Eamon Magee
222 Lower Kilmacud Road,
Goatstown, Dublin 12
Tel: 01-298 7611 &
087 254 9033
E-mail: eamonmagee222@
gmail.com

An Beachaire – The Irish Beekeeper the monthly organ of FIBKA, subscription €30.00 Euro or £25 Sterling for Northern Ireland/Great Britain, post free.

Mr Dermot O Flaherty, Rosbeg, Westport, Co Mayo
Tel No 098-26585 and 087-2464045
E-mail: manager.beachaire@irishbeekeeping.ie

LIBRARY

The library is owned and controlled by FIBKA. It contains very many valuable books ancient and modern, available to members for return postage only.
The Librarian is Mary Montaut , Editor An Beachaire.
Tel: 01 286 0497
Email: editor.beachaire@irishbeekeeping.ie t

EDUCATION

The Federation of Irish Beekeepers' Associations (FIBKA) examination system is run by the Education Officer Michael Maunsell under the direction of the Examination Board; the Board is made up of members from FIBKA and the Ulster Beekeepers' Association (UBKA) who are appointed by the Executive Council.
There are seven levels of examination:Preliminary, Intermediate, Intermediate Apiary practical, Practical Beemaster, Senior, Lecturer and Honey Judge
Preliminary, Intermediate,Senior, Lecturer and Honey Judge Examinations are held during the SummerCourse at Gormanston and Preliminary and Intermediate and Senior examinations are also held at Provincial Centres. In the case of the Intermediate and Senior provincial examinations they alternate between Practical (odd years) and Scientific (even years).
The Lecturer's examination takes place in the presence of three Examiners, one of whom is the invited Gormanston Summer Course lecturer and also acts as the Extern Examiner.

The Intermediate Proficiency Apiary Practical Examination, the Practical Beemaster Examination and the Apiary Practical component of the Senior Examination are arranged by the Education Officer and take place in the candidate's own apiary during the beekeeping season and are conducted by two Examiners.
The seven levels of examinations for proficiency certificates and their eligibility requirements are as follows:

1. Preliminary:
For beginners - no prerequisites.
2. Intermediate:
The Preliminary Certificate of the FIBKA or the BBKA Basic Certificate must be held for at least one year.
3. Intermediate Proficiency Apiary Practical
The Intermediate Proficiency Apiary Practical Examination is intended to be part of a stream that will lead to the Practical Beemaster Certificate. The examination is designed to be less "academic" and there are no written examination papers.
The examination will take place in the candidate's own apiary and the Examiners will be two Federation Lecturers appointed by the Education Officer. The pass mark is 70%.
The prerequisites for Intermediate Proficiency Apiary PracticalExamination are: the Preliminary Certificate and at least three years' beekeeping experience satisfactory to the Education Board.
4. Practical Beemaster:
The prerequisites for the Practical Beemaster Certificate are the Preliminary Certificate, the Intermediate Apiary Practical and at least five years' beekeeping experience satisfactory to the Examination Board.
5. Honey Judge:
Intermediate and Practical Beemaster Certificates, successful showing, having obtained a minimum of 200 points at major shows and a record of stewarding under at least four FIBKA Honey Judges.
6. Senior:
Intermediate Certificate and at least five years beekeeping experience. There are four elements to this examination; Scientific written, Practical written, Microscopy practical and Senior Apiary practical.
7. Lecturer:
Senior Certificate.

Provincial Examinations
Preliminary, Intermediate and Senior examinations will be held at provincial centres on the Saturday closest to 6th April(Intermediate) and May 24th (Preliminary). Please note that the minimum number of candidates for a Centre is five for Intermediate and Senior and ten for Preliminary. Neighbouring associations may pool their candidates to reach those numbers. A candidate may sit one paper at the Provincial Examination and the other paper at the Summer Course in Gormanston or both papers in Gormanston.

The fees for all examinations are valid for the year of application only and are listed on the application forms which may be downloaded from the website. In extreme cases, such as illness (a doctor's certificate must be provided); the examination fee may be held over for one year. There are separate entry forms for the Provincial and Gormanston Summer School Examinations

Fees for Repeat Examinations are the same as for the original examination. Applications to sit the Examinations should be sent to the Education Officer, before the closing dates given above for the Provincial Examinations (applications are however acceptable up to one week after the closing date on payment of a late entry fee which is equal to double the original fee) and before May 1st for the Summer Course Examinations Applications for the Preliminary Examination are also accepted at the Summer Course.

NATIONAL HONEY SHOW

This is held at Gormanston College in conjunction with the annual Beekeeping Course. The Schedule contains 41 Open Classes and 3 Confined classes with €1,000 in prizes. Over 30 Challenge Cups and Trophies are presented for the competition.

Honey Show Secretary: Mr Graham Hall
Weston,38 Elton Park, Sandycove, Co. Dublin.
Tel: 01-2803053 & 087-2406198
E-mail: grahamhall@outlook.ie

INSURANCE

The limit of indemnity of public liability policy is €6.500, 000 arising from one accident or series of accidents. There is also product liability of €6.500, 000 arising from any one claim.

The policy extends to all registered affiliated members whose subscriptions are fully paid up and whose name is entered in the FIBKA register held by the Treasurer.

FIBKA

ASSOCIATION SECRETARIES

ARMAGH & MONAGH
Mrs. Joanna McGlaughlin
26 Leck Road,
Stewartstown Co Tyrone
BT71 5LS
Tel No 048-87738702/077-68107984.
secretary@ambka.org

Ashford
Ms Michele O'Connor,
087 2505205
info@wicklowbees.com

Ballyhaunis
Mr Gerry O'Neill,
Drimineen South,
Knock Road, Claremorris,
Co Mayo.
Tel No 087 2553533
ballyhaunisbeekeepers@gmail.com

Banner
Mr Frank Considine,
Clohanmore Cree,
Kilrush, Co Clare,
Tel No 087-6740462,
bannerbees@gmail.com ,

Beaufort
Mr Padruig O'Sullivan,
Beaufort Bar & Restaurant,
Beaufort, Co Kerry.
Tel No 087-258993006,
beaurest@eircom.net

Carbery
Mr Sean O'Donovan,
Drominidy, Drimoleague,
Co Cork.
Tel No 087-7715001.
seanodonovan10@gmail.com

Co Cavan
Mr Alan Brady,
Shanakiel House,
Drumnagran, Tullyvin,
Co Cavan
Tel No 086-8127920
alan@alanbrady.ie or Info@alanbradyelectrical.com

Co Cork
Mr Robert McCutcheon,
Clancoolemore, Bandon,
Co Cork.
Tel No 023-8841714.
bob@cocorkbka.org

Co Donegal
Mr Dan Thompson,
Highfield, Loughnagin,
Letterkenny, Co Donegal
Tel No 074-9125894
dthompson@eircom.net

Co Dublin
Mr Liam McGarry,
24 Quinn's Road,
Shankill, Co. Dublin
Tel No 087 2643492.
mcgarryliam@gmail.com

Co Galway
Dr Anna Jeffrey Gibson,
Ballyclery, Kinvara,
Co Galway
secretary@galwaybeekeepers.com

Co Kerry
Mr Ruary Rudd,
Westgate, Waterville,
Co Kerry.
Tel No 066-9474251.
rrudd@eircom.net

Co Limerick
Mr Gus McCoy,
Mount Catherine Clonlara
Co. Clare
Tel No 087 1390039 :
gusmccoy1@eircom.net

Co Louth
Mr Tom Shaw,
201 Ard Easmuinn, Dundalk,
Co Louth
Tel No 042-9339619/086-2361286,
tshaw@iol.ie

Co Longford
Mr Joe McEntegart,
Cleanrath, Aughnacliffe,
Co Longford.
Tel No 087-2481340.
josephmcentegart@yahoo.com

Co Mayo
Ms Helen Thompson,
Graffy, Killasser,
Swinford, Co. Mayo.
Tel No 087-7584835
info@mayobeekeepers.com or helen.mmooney@gmail.com

Co Offaly
Mrs Geraldine Byrne,
4 Sheena, Charleville Rd,
Tullamore, Co Offaly
Tel 086-3464545,
loureiro.byrne@gmail.com

Co Waterford
Ms Colette O'Connell,
4 Davis Street, Dungarvan,
Co Waterford
Tel No 058-41910,
coletteoconnell@ymail.com

Co Wexford
Mr John Cloney,
Ballymotey Beg,
Enniscorthy, Co.Wexford.
Tel No 087 9801015
countywexfordbeekeepers@gmail.com

Chorca Dhuibhne
Ms Juli Ni Mhaoileoin,
Burnham, Dingle,
Co Kerry
Tel No 086-8337733,
julimaloneconnolly@
gmail.com

Chonamara
Mr Billy Gilmore,
Maam West, Leenane,
Co. Galway
Tel No 091-571183/087-
7942028,
b.gilmore@
connemarabeekeepers.ie

Digges & Dist
Mr Walter Sharpley,
Aughayoula, Ballinamore
Co. Leitrim
Tel No 086 1236207
waltersharpley@gmail.com

Duhallow
Mr Andrew Bourke,
Pallas, Lombardstown,
Mallow, Co Cork
Tel No 087-2783807.
bourke.andy@gmail.com

Dunamaise
Mr Thomas Hussey,
Glenside Portlaoise,
Co. Laois
thomasjhussey@eircom.net

Dunmanway
Elke Hasner,
Kilnarovanagh, Toames,
Macroom, Co.Cork
026 46312/ 087 2525771
elkehasner@gmail.com

East Cork
Mrs Bridie Terry,
Ait na Greine, Coolbay,
Cloyne, Co Cork.
Tel No 021-4652141.
aitnagreine@gmail.com

East Waterford
Mr Michael Hughes,
51 Woodlawn Grove, Cork
Road, Waterford
Tel No 051-373461.
waterfordbees@gmail.com

Fingal
Mr John McMullan,
34 Ard na Mara Crescent,
Malahide, Co Dublin
Tel No 01-8450193.
jmcmullan@eircom.net

Foyle
Mr Martin Coleman,
Greencastle,
Co. Donegal
foylebeekeepers@
gmail.com

Gorey
C/O President,
Gerard M Williams,
Carrigbeg,
Gorey Co. Wexford
Tel No 053-9421823/086-
3634134 e-mail
geraldandvera@eircom.net

Inishowen
Mr Paddy McDonagh,
Milltown, Carndonagh,
Co Donegal.
Tel No 074-9374881.
paddymcdonough@eircom.
net

Iveragh
Mr Shannon
Ware, 4 Ballinskelligs
Holiday Homes,
Ballinskelligs,
County Kerry.
Tel No: 083-3862345
research@gamelab.ca

Killorglin
Mr Declan Evans,
Reeks View Lodge,
Killorglin, Co Kerry.
Tel: 087 175 4078, :
declanjevans@gmail.com

Kilternan
Ms Mary Montaut,
4 Mount Pleasant Villas,
Bray, Co Wicklow.
Tel No 01-2860497.
mmontaut@iol.ie

Mid-Kilkenny
Jer Keohane,
Jenkinstown Park, Kilkenny.
Tel. 056-7767195 / 087-
2523265 /
jkeohane@iece.ie

New Ross
Mr Seamus Kennedy,
Churchtown,
Feathard-on Sea,
New Ross, Wexford
Tel No 051-397259/
086- 3204236.
seamus.kennedy@yahoo.
co.uk

North Cork
Mr. Eamon Nelligan
3 Carriagroghera
Fermoy Co. Cork.
ciaranneligan@gmail.com

North Kildare
Mr Norman Camier,
34 Lansdowne Park,
Templeogue, Dublin 16.
Tel No 01-4932977/
087-2848938,
norman.camier@gmail.com

Nth Tipperary
Mr Jim Ryan,
"Innisfail", Kickham Street,
Thurles, Co Tipperary.
Tel No 0504-22228.
jimbee1@eircom.net

FIBKA

Roundwood
Mr John Coleman,
Hillside Cottage,
Roundhill Haven, Clara
Beg, Roundwood, Co.
Wicklow Tel No 087-795
4385, colemanjkc@gmail.
com

Sliabh Luachra
Mr Billy O'Rourke,
Dooneen, Castleisland,
Co Kerry
Tel No 066-7141870,
siobhancorourke@eircom.
net

Sligo/Leitrim
Mr Peter Carter,
Doon West, Gurteen,
Co.Sligo
slbasecretary@gmail.com

Sneem
Mr Frank Wallace,
Boolananave, Sneem.
County Kerry.
Tel No 086 3522205,
franksneem@hotmail.com

South Donegal
Mr Derek Byrne,
Carrick West, Laghey,
Co Donegal.
Tel No 074-9722340.
dcbyrne@eircom.ie

South Kildare
Mr Liam Nolan,
Newtown, Bagnelstown,
Co Carlow.
Tel No 059-9727281.
liamnolannt@gmail.com

Sth Kilkenny
Mr John Langton,
Coolrainey,
Graiguemanagh,
Co Kilkenny
Tel No 086-1089652,
jjlangton@eircom.net

Sth Tipperary
Mr P J Fegan,
Tickinor, Clonmel,
Co Tipperary.
Tel No 086 1089652,
feganpj@eircom.net

Sth West Cork
Ms Gobnait O'Donovan,
38 McCurtain Hill,
Clonakilty, Co Cork.
Tel No 023-
8833416/083-3069797
gobnaitodonovan@gmail.
com

Sth Wexford
Mr. Dermot O'Grady,
Linden House, Horetown
North, Foulksmills, Co.
Wexford Tel: (051) 565651,
dermaloid@gmail.com

Suck Valley
Ms Anne Towers,
Doonwood,
Mount Bellew, Co Galway.
Tel No 0909-684547/
087-6305714,
annevtravers@gmail.com

The Kingdom
Ms Rebecca Coffey,
75 Ashgrove, Tralee,
Co Kerry
Tel No 066- 7169554,
bexk8@yahoo.co.uk

The Royal Co
Ms Geraldine McCann,
Mooretown, Ratoath,
Co. Meath.
geraldine.toole@ucd.ie

The Tribes
Mr Eoghan O'Riordan,
28 Arbutus Avenue,
Renmore, Galway.
Tel No 091-753470/
087-6184132,
landservices@eircom.net

West Cork
Ms Jacqueline Glisson,
Costa Maningi, Derrymihane
East, Castletownbere,
Co Cork. Tel No 086-
3638249,
jglisson@eircom.net

Westport
Mr Dermot O Flaherty,
Rosbeg, Westport,
Co Mayo
Tel No 098 26585/
087-2464045,
info@mayo-westport.com

INTERNATIONAL BEE RESEARCH ASSOCIATION WEB http://www.ibra.org.uk

The International Bee Research Association (IBRA) promotes the value of bees by providing information on bee science and beekeeping worldwide. This Registered Charity was founded in 1949 and is supported by members from around the world. IBRA publishes books, journals and other information and organises conferences on all species of bees, beekeeping and bee conservation. IBRA has one of the largest international collections of bee books and journals, as well as the Eva Crane / IBRA historical collection and a photographic collection.

PUBLICATIONS

Journal of Apicultural Research

A peer reviewed scientific journal that is worldwide and world class. JAR is edited by Senior Editor Norman Carreck and our team of editors in Argentina, Germany, Greece, Switzerland, Turkey and the USA. It is published for IBRA by Taylor & Francis and available both print and online. Published five times a year it contains the latest high quality original research from around the world, covering all aspects of the biology, ecology, natural history, culture and conservation of all types of bee.

Bee World

IBRA's quarterly international popular journal provides a world view on bees and beekeeping. It is edited by Kirsten Traynor in the USA and published for IBRA by Taylor & Francis and available both print and online. It covers all topics from bee history to the latest findings in bee science in a digestible form.

SCIENCE DIRECTOR
Norman Carreck
Laboratory of Apiculture and Social Insects,
University of Sussex
Falmer
Brighton
East Sussex
BN1 9QG
01273 872587
norman.carreck@btinternet.com

SECRETARY
Ivor Davis, 91,
Brinsea Road,
Congresbury, Bristol,
S49 5JJ, UK.
davisi@ibra.org.uk

Correspondence to:
mail@ibra.org.uk

www.ibra.org.uk
https://www.facebook.com/IBRAssociation
https://twitter.com/IBRA_Bee

The IBRA BOOKSHOP
Through our online bookshop IBRA sells IBRA publications, together with a wide range of other publications at competitive prices as well as posters, gifts, DVD's and sundries. Our bookshop also visits the BBKA Spring Convention, the National Honey Show and other events for you to browse before you buy and purchase items post free.
http://ibrabee.org.uk/index.php/our-shop

MEMBERSHIP
IBRA is a truly international organisation, with members all over the world. IBRA Members receive Bee World free of charge, together with discounts on IBRA publications and other benefits. For details on current membership rates and how to join please visit our website.

The International Bee Research Association is a Company limited by Guarantee,

Registered in England and Wales, Reg. No. 463819,
Registered Office: 91 Brinsea Road, Congresbury, Bristol, BS49 5JJ, UK and is a Registered Charity No. 209222.

Information about all IBRA publications and services can be found via our web site: www.ibra.org.uk

Available and published by IBRA but also available from Northern Bee Books at www.northernbeebooks.co.uk

THE INSTITUTE OF NORTHERN IRELAND BEEKEEPERS (INIB)

www.inibeekeepers.com

Annual Conference and Honey Show. 22nd October 2016
Speakers: Sue Cobey , Harry Owens, Gregg Fariss and Timothy Lawrence
Lough Neagh Discovery Discovery centre, Oxford Island, Lurgan, BT66 6NJ

Objectives of the Institute

The Institute is established to advance the service of apiculture and to promote and foster the education of the people of Northern Ireland and surrounding environs without distinction of age, gender, disability, sexual orientation, nationality, ethnic identity, political or religious opinion, by associating the statutory authorities, community and voluntary organisations and the inhabitants in a common effort to advance education, and in particular:

- to raise awareness amongst the beneficiaries about bees, bee-keeping and methods of management;
- to foster an atmosphere of mutual support among bee-keepers and to encourage the sharing of information and provision of helpful assistance amongst each other.

Affiliation

INIB is affiliated to the British Beekeepers Association.
With 21,100 members the British Beekeepers Association (BBKA) is the leading organisation representing beekeepers within the UK.
As an INIB member, affiliation gives the following benefits.

- BBKA News
- Public Liability Insurance
- Product Liability Insurance
- Bee Disease Insurance available
- Free Information Leaflets to Download
- Members Password Protected Area and Discussion Forum
- Correspondence Courses
- Examination and Assessment Programme
- Telephone Information
- Research Support
- Legal advice
- Representation and lobbying of Government, EU and official bodies.

Events

Events

The Institute holds an annual conference and honey show. The Institute brings to Northern Ireland world renowned expert speakers from USA and Europe to give talks to beekeepers on the latest research and up to date beekeeping methods.

Education

Demonstrations on various topics such as mead making, preparing honey for shows are held during the year.

MEMBERSHIP SECRETARY
Lyndon Wortley
Teemore Grange
224 Marlacoo Rd,
Portadown,
BT62 3TD
Membershipsecretary@
inibeekeepers.com

CHAIRMAN
Michael Young MBE
101 Carnreagh,
Hillsborough
BT26 6LJ
02892689724
chairman@
inibeekeepers.com

Holders of the Institute of Northern Ireland Beekeepers Honey Judge Certificate

001.	MICHAEL BADGER MBE	01132 945879	buzz.buzz@ntlworld.com
002.	GAIL ORR	02892 638363	gail.orr@belfasttrust.hscni.net
003.	CECIL MCMULLAN	02892 638675	Madeline.mcmullan@hotmail.co.uk
004.	HUGH MCBRIDE	02825 640872	lorraine.mcbride@care4free.net
005.	LORRAINE MC BRIDE	02825 640872	lorraine.mcbride@care4free.net
006.	BILLY DOUGLAS	02897 562926	
007.	MICHAEL YOUNG MBE	02892 689724	chairman@ inibeekeepers.com
008.	FRANCIS CAPENER	01303 254579	francis@honeyshow.freeserve.co.uk
009.	MARGARET DAVIES	01202 526077	marg@jdavies.freeserve.co.uk
010.	IAN CRAIG	01505 322684	ian'at'iancraig.wanadoo.co.uk
011.	DINAH SWEET	02920 756483	
013.	LESLIE M WEBSTER	01466 771351	leswebster@microgram.co.uk
014.	REDMOND WILLIAMS	003535242617	emwilliams@eircom.net
015.	TERRY ASHLEY	01270 760757	terry.ashley@fera.gsi.gov.uk
016.	IVOR FLATMAN	01924 257089	ivorflatman@supanet.com
017.	ALAN WOODWARD	001302 868169	janet.woodward@virgin.net
018.	DENNIS ATKINSON	01995 602058	dhmatkinson@tesco.net
019	LEO MCGUINNESS	028711 811043	pmcguinness@glendermott.com
020	TOM CANNING		tjcanning@btinternet.com
023	ALAN BROWN	01977 776193	alanhoneybees4u@talktalk.net
024	DAVID SHANNON	01302772837	dave_aca@tiscali.co.uk
USA			
021	ROBERT BREWER		rbrewer@arches.uga.edu
022	BOB COLE		
023	ANN HARMAN		

LABORATORY OF APICULTURE & SOCIAL INSECTS (LASI)

UNIVERSITY OF SUSSEX

FURTHER INFORMATION CONTACT
Francis L. W. Ratnieks,
Professor of Apiculture
Laboratory of Apiculture & Social Insects (LASI)
Department of Biological & Environmental Science
University of Sussex, Falmer, Brighton BN1 9QG, UK

01273 872954 (landline),
07766270434 (mob)
F.Ratnieks@Sussex.ac.uk
www.sussex.ac.uk/lasi

Youtube:
LASI Bee Research & Outreach

LASI was founded in 1995 and is headed by Francis Ratnieks, who is the UK's only Professor of Apiculture. Prof. Ratnieks received his training in honey bee biology at Cornell University and the University of California in the USA. Whilst in the USA he was also a part-time commercial beekeeper with up to 180 hives used for almond pollination and comb honey production.

From 1995 to 2007, LASI was based at the University of Sheffield. In 2008 Prof. Ratnieks moved to the University of Sussex, which provided LASI with excellent facilities for honey bee research. There is an integrated lab space and offices sufficient for 13 researchers with an adjoining apiary, garden, equipment shed and workshop. There are further apiaries on the university campus and in the surrounding countryside.

LASI is the largest university-based laboratory studying honey bees in the UK and is set up both to undertake research and to train the next generation of honey bee scientists. Undergraduate students receive lectures on honey bee biology and can also do research projects on honey bee biology in their final year and assist LASI research via summer bursaries. Graduate students take a PhD that focuses in a particular area of research. Postdoctoral researchers can learn new skills to complement the training they received during their PhD.

LASI research focuses on both basic and applied questions in bee biology and beekeeping. Basic research areas include communication, foraging, colony organization, nestmate recognition and guarding, and conflict resolution. Applied research areas include improved beekeeping techniques, studies of bee foraging and the value of different plants for honey bees and other pollinators, crop pollination, practical

studies of honey bee diseases and their management, and practical measures for bee conservation. Collectively, the LASI team has 80 years of research experience with honey bees.

As well as research and teaching, LASI places great emphasis on outreach and communication. Each year LASI runs workshops, gives talks and writes many outreach articles so that research results are also transferred to beekeepers, gardeners, farmers, land owners, the media, the general public, and policy makers.

NB: This is the last entry received and the publisher cannot be sure if this is accurate.

THE NATIONAL DIPLOMA IN BEEKEEPING

The Examinations Board for the National Diploma in Beekeeping was set up in 1954 to meet a need for a beekeeping qualification above the level of the highest certificate awarded by the British, Scottish, Welsh and Ulster Associations.

The Diploma Examination, as designed by the Board, was considered to be an appropriate qualification for a County Beekeeping Lecturer or a specialist appointment requiring a high level of academic and practical ability in beekeeping. It is the highest beekeeping qualification recognised in the British Isles and a high percentage of the past and present holders of the Diploma have given distinguished service to beekeeping education at all levels.

Although the post of County Beekeeping Lecturer has now disappeared, this has merely emphasised the need for some beekeepers to face the challenge of this examination and maintain the high level skills and knowledge needed to keep pace with the increased problems facing all beekeepers at the present time.

The Board consists of representatives from a wide range of organisations and from Government Departments and together form an impressive amalgam of expert knowledge in Beekeeping and Education. Although the National Beekeeping Associations are represented on the Board it is entirely independent of them.

Normally the highest certificate of one of the National Associations is a necessary criterion for eligibility to take the Examination for the Diploma which is held in alternate years. The Written Examination is taken in March, and the Practical, in three sections plus a viva-voce is held in later in the same year.

The Board also organises an annual Advanced Beekeeping Course covering various parts of the syllabus that are difficult to cover by independent study. Lasting

HON. SECRETARY
Mrs Margaret Thomas NDB
Tig na Bruaich,
Taybridge Terrace,
Aberfeldy, Perthshire,
PH15 2BS.

CHAIRMAN,
Ivor Davis NDB
91 Brinsea Road,,
Congresbury
BS49 5JJ
07831 379222

a working week, they cover the main sections of the Syllabus and represent the highest level of training available to British Beekeepers at the present time. The outside lecturers are each acknowledged experts in their particular field.

In addition the Board organize various short courses at locations in the UK on a number of topics.

For further details regarding the Diploma write, enclosing a stamped A4 SAE to the Secretary, or visit our website: http://www.national-diploma-bees.org.uk/

Those who have gained the National Diploma in Beekeeping

Matthew Allan
*Harry Allen
*Harrison Ashforth
*John Ashton
Dianne Askquith-Ellis
David Aston
*John Atkinson
*Miss E.E. Avey
Dan Basterfield
Ken Basterfield
Bridget Beattie
*Brig. H.T. Bell
R.W. Brooke
Norman Carreck
*Rosina Clark
Charles Collins
Gerry Collins
*Tom Collins
*Robert Couston
John Cowan
S. J. Cox
Jim Crundwell
Beulah Cullen
Celia Davis
Ivor Davis
*Alec S.C. Deans
Clive De Bruyn
A.P. Draycott
M. Feeley
*Barry Fletcher
* David Frimston
Oonagh Gabriel
George Gill
*Reg Gove
*Eric Greenwood
Pam Gregory
Anthony R.W. Griffin
* Robert Hammond
Ben Harden
Tony Harris
C.A. Harwood
*Leslie Hender
*Alf Hebden
*Ted Hooper MBE
Geoff Hopkinson BEM
*G. Howatson
* Geoff Ingold
George Jenner
C. F. Jesson
Simon Jones
A.C. Kessel
W.E.Large
Adam Leitch
G.W. Lumsden
*Henry Luxton
A.S. Mcclymont
J.I. Macgregor
Ian Mclean
Ian A. Maxwell
Paul Metcalf
J.Mills
*Bernhard Mobus
Margaret Murdin
G. N'Tonga
*Peter Oldrieve
Gillian Partridge
* E.H. Pee
I.E. Perera
E.R. Poole
Bill Reynolds
Pat Rich
*Fred Richards
E. Roberts
*Arthur Rolt
*Jeff Rounce
Graham Royle
J. Ryding
J.H. Savage
*Donald Sims
F.G. Smith
*George Smith
J.H.F. Smith
Robert Smith
*Ken Stevens
*J. Swarbrick
Margaret Thomas
Adrian Waring
Alastair Welch
Brian Welch
J. Wilbraham

* - deceased

THE NATIONAL HONEY SHOW

www.honeyshow.co.uk

26TH – 28TH OCTOBER 2017.

This venue is excellent with Free car parking
Just off the M25 junction 10
Rail from Waterloo to Esher

The Show itself is a wonderful competitive exhibition of all the products of the bee-hive, coupled with an excellent series of lectures, workshops and a wide variety of trade and educational stands.

We recommend that you attend all three days, and suggest that you become a member of the Show – just **£15.00** per annum

For further information, please write to the Hon General Secretary, or Email: showsec@zbee.com or visit our website www.honeyshow.co.uk

- The National Honey Show is the premier honey show within the United Kingdom.
- Although it is named the "National Honey Show", it includes a strong international element.
- As well as the competitive content of the Show, there is also a full programme of lectures and workshops.
- In the Sales Hall, all the major traders and educational organisations are present.
- Further information is readily available on the website www.honeyshow.co.uk or from the Hon General Secretary showsec@zbee.com

HON SECRETARY
J.D. Hendrie
26 Coldharbour Lane
Hildenborough
Tonbridge
TN11 9JT
01732 833894

✉ ☎

The Native Irish Honey Bee Society
Apis mellifera mellifera

NATIVE IRISH HONEY BEE SOCIETY

CHAIRPERSON:
Mr. Gerard Coyne
Chairperson@nihbs.org

HON. SECRETARY:
Dympna Summerville
Secretary@nihbs.org

TREASURER;
Trevor Gould
treasurer@nihbs.org

PUBLIC RELATIONS OFFICER:
Pat Deasy
pro@nihbs.org

WEBMASTER:
Mr. Jonathan Getty
webmaster@nihbs.org
www.nihbs.org

FACEBOOK PAGE
www.facebook.com/native-irish-honey bee-society

What is the Native Irish Honey Bee Society?
NIHBS was established in November 2012 by a group of beekeepers who wish to support the various strains of Native Irish Honey Bee (Apis mellifera mellifera) throughout the country. It is a cross border organisation and is open to all. It consists of members and representatives from all corners of the island of Ireland.

Aims and Objectives -
To promote the conservation, study, improvement and re-introduction of Apis mellifera mellifera (Native Irish Honey Bee), throughout the island of Ireland.

- To establish areas of conservation throughout the island for the conservation of the Native Honey Bee.
- To promote formation of bee improvement groups.
- To provide education on bee improvement and to increase public awareness of the native honey bee.
- To act in an advisory capacity to groups and individuals who wish to promote it.
- To co-operate with other beekeeping organisations with similar aims.
- To seek the help of the scientific community and other stake holders in achieving our aims and objectives.

NIHBS

The Four Seasons journal is issued to members quarterly.
It contains articles of scientific and practical interest to beekeepers.

Regional groups provide a link between NIHBS members and committee. They organise Open days and workshops.

NIHBS - Plans for the future

- Promote the conservation of the native honeybee in Ireland.
- Continue to part fund and support research:
    - NUI Galway - Development of a breeding programme for Native Irish Honeybees *Apis mellifera mellifera* with natural tolerance/resistance to *Varroa destructor*.
    - Limerick Institute of Technology – Genetic Analysis of the Native Irish Honeybee, *Apis mellifera mellifera*.
- Disseminate information on programme findings to beekeepers and public.
- Teach queen rearing skills at NIHBS Workshops and Open days.
- Regional groups.
- Promote voluntary conservation areas.
- Talks and lectures
- Annual conference

Why Join NIHBS?

- Information on beekeeping events around Ireland – North and South
- queen rearing workshops, talks and lectures.
- Information on how to obtain Native Honey Bees
- Conference discounts
- Discounted entrance fees to events run by NIHBS
- Eligibility to schemes coordinated by NIHBS
- A network of beekeepers interested in our native honeybee

Membership of NIHBS costs 25 Euro or 25 Pounds Sterling.
Join NIHBS on website: http://nihbs.org

ROTHAMSTED RESEARCH

www.rothamsted.ac.uk

ROTHAMSTED RESEARCH
Department of AgroEcology,
Rothamsted Research
Harpenden,
Hertfordshire.
AL5 2JQ

STAFF
DR ALISON HAUGHTON
DR JASON LIM
DR SAMANTHA COOK
JENNY SWAIN
THOMAS DAVID (PHD STUDENT)
STEVE KENNEDY (BEEKEEPER)

The Rothamsted site provides a unique working environment with specialist modern equipment facilitating research on plant and microbial metabolites, molecular biology and synthetic and analytical chemistry. There is an experimental farm for complex field experiments, and there is a suite of glasshouses, controlled environment facilities, an insectary and a state-of-the-art bioimaging suite housing three new electron microscopes and a confocal laser scanning microscope. Experimental design and analysis are backed up by excellent statistical, computing and library support.

BEE BEHAVIOUR AND POLLINATION ECOLOGY

We are investigating the interaction between bees, crops and the agricultural environment. The spatial and temporal foraging behaviour of honey bees and bumble bees within agricultural areas is being compared. Harmonic radar is being used to track flying bees, and other pollinators such as butterflies, to obtain new information about their flight paths, forage ranges, food preferences and orientation mechanisms.

An integrated model for predicting bumblebee population success and pollination services in agro-ecosystems will be developed by Rothamsted and colleagues at the Environment & Sustainability Institute at the University of Exeter and the University of Sussex, and will provide a powerful tool for shaping recommendations for land managers and policy makers for the sustainable spatial management of pollination within arable and horticultural production systems.

Various qualities of different varieties of crops (oilseed rape and short rotation coppice willows) as important resources for bees are being investigated. The nutritional

value of the nectars and pollens, effects on bee fitness and behaviour are key areas of interest.

HONEY BEE PATHOLOGY

Rothamsted's research on the natural history and epidemiology of the infections and parasites of bees has had wide international recognition. However, research on honey bee pathology is currently suspended due to changes in funding available from Defra for bee health. Over the last 20 years, this work focused on *Varroa destructor* and the losses caused by honey bee virus infections that the mite transmits. In a collaborative project with Horticulture Research International (at University of Warwick), investigating potential biological control agents of *V. destructor*, the research identified and characterised fungal pathogens which are active against the mite but which are relatively safe for bees and other beneficial insects. Biological control offers an environmentally acceptable approach to the problem that could have considerable economic benefits, and we are actively seeking funding to continue this work.

We are currently analysing data from an Insect Pollinator Initiative funded project that assessed the impact of emergent diseases, including the Varroa associated Deformed wing virus, and the Microsporidian *Nosema ceranae* on the flight performance and orientation ability of honeybees and bumblebees and its consequences for bee populations.

HARMONIC RADAR

The use of harmonic radar in insect behaviour studies has been pioneered at Rothamsted. A transponder weighing just a few milligrams fitted to the thorax of bees picks up the interrogation radar signal and immediately emits a signal at a different frequency, which is then received by the radar. A recently awarded European Research Council grant will now enable cutting-edge development of the harmonic radar to allow us to collect data for entire adult life-spans and foraging ranges for multiple individuals of bee species, thus allowing us whole new insights into bee behaviour and pollination ecology.

INFORMATION EXCHANGE

Expertise in bee research is drawn upon by scientific colleagues world-wide and there are research links with institutes and universities in this country and abroad. Research findings are published in scientific journals but popular articles are also written for the beekeeping and agricultural press. Effective communication of our science by staff members is delivered via a vigorous programme of lectures presenting to national and local beekeeping associations and participation in various public media, including BBC programmes.

FUNDING

Rothamsted receives funds for research from the Biotechnology and Biological Sciences Research Council, through competitions and contracts from the Department for Environment, Food and Rural Affairs, the European Community, from Levy boards, commercial and other organisations. The support of the bee research programme in recent years by grants from the British Beekeepers Association, C. B. Dennis British Beekeepers Research Trust, the Eastern Association of Beekeepers and the Bedfordshire, Cambridgeshire, Norfolk, St Albans and Hertfordshire and High Wycombe Beekeepers Associations is gratefully acknowledged.

For more information visit: **http://www.rothamsted.ac.uk**

THE SCOTTISH BEEKEEPERS' ASSOCIATION

The Scottish Beekeepers' Association is a Scottish Charitable Incorporated Organisation registered in Scotland, number SCI009345

Purposes of the Association

The organisation's purposes are to support honeybees and beekeepers, to improve the standard of beekeeping, and to promote honeybee products in Scotland through:

The advancement of education in relation to the craft of beekeeping; The advancement of the heritage, culture and science of beekeeping; and the advancement of environmental protection by conservation of the honeybee.

The SBA arranges courses and awards certificates to successful candidates in a comprehensive education system. It also actively promotes beekeeping by informing the public, including the young, about bees and their benefits to the environment.

INSURANCE AND THE COMPENSATION SCHEME

All members of the SBA have insurance against Public Liability. The SBA Compensation Scheme is restricted to bee colonies located in Scotland and allocates part-replacement value for damage by vandalism, fire, theft and certain brood diseases.

LIBRARY

The SBA Moir Library in Edinburgh has one of the world's finest collections of beekeeping books. A library card is issued annually to every member who can borrow books at the cost of return postage only. Details may be obtained from the Library Officer.

MARKETS

Advice is given on all aspects of marketing honey products at appropriate times. Suggested bulk, wholesale and retail prices are notified in the magazine.

GENERAL SECRETARY
Dr John Wilkinson
Oakendean Lodge
Melrose
TD6 9HA
01896820500
secretary@
scottishbeekeepers.org.uk

HON PRESIDENT
Vacant

HON. VICE PRES,
Iain F Steven
4 Craigie View
Perth
PH2 0DP
01738 621100

Ian Craig
30 Burnside Ave, Brookfield,
Johnstone, Renfrewshire,
PA5 8UT
01505 322684
beekeeper30@btinternet.
com

HON. LIBRARIAN
Mrs. Margaret M. Sharp
City Librarian, City Library
George IV Bridge, Edinburgh

SBA

HON. LEGAL ADVISER,
Taggert, Meil & Mathers
20 Bon Accord Sq,
Aberdeen
01224 588020

INDEPENDENT EXAMINER
Mhairi Callander,
10 Ardross Street,
Inverness,
IV3 5NS

PUBLICATIONS
The Scottish Beekeeper is published monthly and sent post free as part of the annual membership fee of £30 payable to the Membership Officer. Introduction to Bees and Beekeeping is £6.00 plus postage and may be obtained from the Advertising and Publicity Officer.

PUBLICITY
Members can purchase the Association tie, lapel badge, car sticker etc. Details may be obtained from the Shows and Publicity Officer.

SHOWS
Two major annual honey shows are held in Scotland. A honey competition and show with educational displays is held at the Royal Highland Show, Ingliston, Edinburgh in June and the Scottish National Honey Show is conducted at the Dundee Food and Flower Festival in September. Other Honey Shows are run in Ayr, Fife, Inverness, Turiff and at many other locations in Scotland as organised by Local Associations.

TRUSTEES

TRUSTEES

MEMBER AFFILIATED BEEKEEPING ASSOCIATIONS AND THEIR SECRETARIES

Aberdeen,
Rosie Crighton
29 Marcus Cresc
Blackburn, Aberdeen
AB21 0SZ
01224 791181
aberdeenbeekeepers@
gmail.com

Arran Bee Group,
W K McNeish
Seafield, Kildonan,
Isle of Arran,
KA27 8SE
01770 820357
wmcnsh@aol.com

Ayr,
Mrs L Baillie
Windyhill Cottage
Uplands Rd, Sundrum
Ayre, KA6 5JU
01292 570659
lbaillie@sundrum.demon.
co.uk

Borders,
Liz Howell
Oatlands, Houndridge,
Kelso
TD5 7QN
01573 470747
kevhwl@aol.com

Caddonfoot,
Mrs Brenda Lambert
1 High Cottage,
Walkerburn
EH43 6AZ
01896 870428
blambert1962@outlook.
com

Covington and Thankerton,
Angus Milner-Brown
Covington House,
Covington Road,
Biggar
ML12 6NE
01899 308024
angus@therathouse.com

Cowal,
Ceci Alderton
2 Stronechrevich
Strachur
Argyll
PA27 8DF
01369 860445
cecis.farmlet@gmail.com

Dingwall,
Sarah Smyth
14 Cromartie Drive
Strathpeffer
IV14 9DB
07736276238
dingwall.beekeeping@
googlemail.com

Dunblane & Stirling,
Fiona Fernie
Greystones Dunira,
By Comrie
PH6 2JZ
01764 679152
secretary@
dunblanebeekeepers.com

Dunfermline & West Fife
Liz Wyatt
Hollytree Lodge,
Muckhart,
Dollar,
FK14 7JW
01259 781214
dwf@fifebeekeepers.co.uk

East Lothian,
Deborah Mackay
5 Goshen Farm
Steading,
Musselburgh, East Lothian
EH21 8JL
0131 665 8939
eastlothianbeekeepers@
googemail.com

East of Scotland,
Colin Smith
The Laundry House,
Ethie, Inverkielor, Arbroath,
DD1 5SP
secretary@
eastofscotlandbeekeepers.
org.uk

Easter Ross,
Colin Ridley
Stirling Cottage,
Lamington Park,
Kildray,
Ross-shire
IV18 0PE
01862 842410
colinr031@googlemail.com

Eastwood,
Robert Gordon
10 Caribar Drive,
Barrhead, Glasgow
G78 1BQ
0141 5716498
Robert.gordon@ntworld.
com

Edinburgh & Midlothian
Gordon Jardine
20 Pentland Grove,
Edinburgh,
EH10 6NR
07703 528801
gordieric@hotmail.com

Fife,
anice Furness
The Dirdale, Boarhills
St. Andrews,
Fife
KY16 8PP
01334 880 469
jcfurness@dirdale.fsnet.co.uk

Fortingall,
Mrs Sharon Martin
Gardeners Cottage,
Grand Tully,
PH15 2EG
01887840407
sharon.x.martin@btinternet.com

Glasgow District,
Mhairi Neill
3 Machan Ave,
Larkhall,
ML9 2HE
01698 881602
glasgowbeeksec@hotmail.co.uk

Helensburgh and District,
Cameron Macallum,
The Old Police Station,
Arrochar, Argyll.
G83 7AA.
Tel. No. 01301702295.
secretary@helensburghbees.com

Inverness,
Helen Macleod,
66 Grigor Drive
Inverness, IV2 4LS

Kilbarchan and District,
Helena Jackson
11 Criagends Avenue
Quarriers Village
PA11 3SQ
helena.jackson72@sky.com
07789711703

Kilmarnock & District,
J. Campbell
North Kilbryde House,
Stewarton, Kilmarnock,
KA3 3EP
01560 482489
john.d.campbell@talktalk.net

Kintyre & Mid Argyll,
Zoe Weir
1 Lagnagorton,
Clachan, Tarbert,
Argyll, PA29 6XW
lagnagortan@aol.com

Lanarkshire
Susan Fotheringham
info@lanarkshirebeekeepers.org.uk

Lochaber,
Sarah Kennedy
Tigh na Feid,
Achintore Road,
Fort William
PH33 6RN
secretary@lochaberbeekeepers.org

Moray,
Tony Harris
Cowiemuir,
Fochabers,
Moray
IV32 7PS
07884496246
tony@moraybeekeepers.org.uk

Mull and Iona,
Helen Howarth
Cul a Mhill
Ardtun
Bunessan
Isle of Mull
PA67 6DH
01681 700724
helen@waltonhowarth.co.uk

Nairn & District,
Ruth Burkhill
2 Cloves Cottage,
Alves, Moray
IV36 2RA
01343 850041
ruthburkhill@gmail.com

Newbattle,
Helen Nelson,
8/1/ Russle Place,
Edinburgh. 0131 467 3616
helen.nel@hotmail.com

North Ayrshire,
Ruth Anderson
07773 776253
northayrshirebeekeepers@gmail.com

Oban & District,
Nigel Mitchell
Barochreal
Kilninver
Oban,
PA34 4UT
01852 316151
nigel@themitchells.co.uk

Olrig and District,
Robin Inglis
Roadside, Skirza
Freswick,
Wick
KW1 4XX
01955 611260
gailinglis@btinternet.com

Orkney, Sue Spence
Alton House,
Berstane Road,
Kikwall,
Orkney
KW15 1NA
01856 873920
bs3920@yahoo.com

SBA

Peebles-shire,
Amanda Clydesdale
20 Kingsmeadows Gardens
Peebles
EH45 9LB
01721 720563
amanda.clydesdale@
btinternet.com

Perth and District,
Brian Clelland
12 Albert Road,
Scone, Perth,
PH2 6QH
07845375298
info@
perthanddistrictbeekeepers.
co.uk

Skye & Lochalsh,
Joe Grimson
1 Riverside Cottage,
Braeintra, Stromeferry,
IV53 8UP
j.grimson@btinternet.com

South of Scotland,
Debbie Park
Crofthead,
Dalswinton,
Dumfries
DG2 0XY
01387740030
d.parke@yahoo.co.uk

Speyside,
Gerry Thompson
Highland House,
Knockando,
Aberlour, Moray,
AB53 7RP
01340810229
gjthom@sky.com

Sunart,
Ardnamurchan,
Moidart and Morvern
Kate Atchley
Anasmara,
Mingarry, Acharacle,
Argyll & Bute
PH36 4XJ
07774807645
bees@kateatchley.co.uk

Sutherland, Sue Steven
Mulberry Croft,
2 East Newport,
Berriedale
Caithness
KW7 6HA
01539 751 245

Western Galloway
Linda Robertson
07825514726
wgbacontact@gmail.com

Western Isles
Martin Johnstone
3 Upper Bayble
Point
Isle of Lewis
HS2 0QH
martinjohnstone1@tiscali.
co.uk

West Linton and District
Dave Stokes
100 Main Street
Roslin
Midlothian
EH25 9LT
0131 4403477
wlbka@live.co.uk

Kemnay Community Group
Beekeepers
David Wilkinson
18 St Ninians
Moneymusk
AB51 7HF
01467651400
secretary@kemnaybees.org

SBA HONEY JUDGES

Introductory notes:
Check with your preferred judge(s) whether thay are happy to travel to your location and any equipment needed.
Judges will expect reasonable travel expenses to be paid and may request an additional fee.
All such practicalities, and any accommodation, should be ageed in advance.

Enid Brown
Milton House, Main
Street, Scotlandwell,
Kinross KY13 9JA
01592 840582 enidbrown6@gmail.com

Mick Canham
Whinhill Farm House,
Nairn IV12 5RF
01667 404314

Ian Craig
30 Burnside Avenue,
Brookfield, Johnstone,
Renfrewshire PA5 8UT
01505 322684 beekeeper30@btinternet.com

Tony Harris
Cowiemuir, Fochabers,
Moray IV32 7PS
07884 496246 tonyharris316@btinternet.
com

Charlie Irwin
55 Lindsaybeg Road,
Chryston,
Glasgow G69 9DW
0141 779 1333 ceirwin@talktalk.net

High Donohoe
7 Grant Road, Banchory
AB31 5UW
01330 823502

P & Mrs C Mathews
4 Annanhill,
Annan,
Dumfries-shire
DG12 6TN
01461 205525

Brenda McLean
Upper Flat, 2 Invererne
Road, Forres IV36 1DZ
01309 676316

Alan Riach
Woodgate, 7 Newlands
Avenue, Bathgate,
Nr Edinburgh
EH48 1EE
01506 653839
alan.riach@which.net

Claude Wilson
Cedarhill, Auchencloch,
Banknock,
Bonnybridge
FK4 1VA
01324 840227

Dr and Mrs Wright
20 Lennox Row,
Edinburgh EH5 5JW
0131 552 3439 bronwright20lr@btinternet.
com

SBA Speakers List

Introductory notes:
Details of speakers and their topics can be found on the secretaries area of the SBA website.

For other talks, please discuss your needs with one or more speakers, many of whom are willing to adapt a talk for a different audience.

Talks are usually supported by slide presentations/powerpoint. Associations may need to provide a projector and screen. Speakers will expect reasonable travel expenses to be paid and some may request an additional fee.

Check with your preferred speaker(s) whether thay are happy to travel to your location. All these practicalities, and any accommodation requirements, should be agreed with the association secretary in advance.

Dianne Barry 29 Cambridge, Gardens. Edinburgh EH6 5DH
0131 476 3506 & 07949 784888
dianne.barry@me.com

David Brown 1 Grigor Gardens, Inverness IV2 4JU
01463 234466
david@bearradh.co.uk

Enid Brown Milton House, Main Street, Scotlandwell, Kinross KY13 9JA
01592 840582
enidbrown6@gmail.com

Ann Chilcott, Piperhill, Nairn IV12 5DS
01667 404606
ann@chilcott.myzen.co.uk

John Coyle Rose Cottage, Burnton, by Kippen, Stirling FK8 3JL
01786 870674 & 07774 266540
info@beekeepinginscotland.co.uk

Ian Craig 30 Burnside Avenue, Brookfield, Johnstone, Renfrewshire PA5 8UT
01505 322684
beekeeper30@btinternet.com

Prof David Evans Biomolecular Sciences, Building, North Haugh, University of St Andrews, St. Andrews, Fife KY16 9ST
01334 463396
d.j.evans@st-andrews.ac.uk

Fiona Highet SASA, Roddinglaw, Road, Edinburgh EH12 9FJ
0131 244 8817
fiona.highet@sasa.gsi.gov.uk

Tony Harris Cowiermuir, Fochabers, Moray IV32 7PS
07884 496 246
tonyharris316@btinternet.com

Charlie Irwin 55 Lindsaybeg Road, Chryston, Glasgow G69 9DW
0141 779 1333
ceirwin@talktalk.net

Phil McAnespie 12 Monument Road, Ayr KA7 2RL
01292 885660
philmcanespie@btinternet.com

Magnus Peterson Balhaldie House, High Street, Dunblane FK15 0ER
01876 822093
magnus.peterson@strath.ac.uk

Dr Gavin Ramsay 14 Redcliffs, Kingoodie, Dundee DD2 5DL
01382 562105 & 07751 142155
gavinramsay@btinternet.com

Bryce Reynard 39 Old Mill Lane, Inverness IV2 3XP
01463 225887
elizabethreynard@btinternet.com

Alan Riach Woodgate, 7 Newlands Avenue, Bathgate, Nr Edinburgh EH48 1EE
01506 653839
alan.riach@which.net

Willie Robson Chain Bridge Honey Farm, Horncliffe, Berwick upon Tweed, TD15 2XT
01289 382362
info@chainbridgehoney.co.uk

Scottish Bee Inspectorate: Steve Sunderland, Lead Bee Inspector
SGAHW, P Spur, Saughton House, Broomhouse Drive, Edinburgh EH11 3XD
0300 244 6672
steve.sunderland@.gsi.gov.uk

Scottish Bee Inspectors
as above 0300 244 6672 bees_mailbox@.gsi.gov.uk as above

Graeme Sharpe (SAC) SAC Consulting, Beekeeping Unit, John Niven Building, Auchincruive Estate, Ayr KA6 5HW
01292 525375
graeme.sharpe@sac.ac.uk

Julian Stanley Hewbank Knowe, Sornhill, Galston, Ayrshire KA4 8NF
01563 821831
julianrstanley@aol.com

Dr Peter Stromberg 21 Woodside, Houston, Renfrewshire PA6 7DD
01505 613830
pstromberg1@aol.com

Margaret Thomas Tighnabraich, Taybridge Terrace, Aberfeldy, Perthshire PH15 2BS
01887 829710
zyzythomas@waitrose.com

John Vendy High Bentham, North Yorkshire, LA2 7BP
07973 907325
John@topbarbeehive.co.uk

Bron Wright 20 Lennox Row, Edinburgh EH5 3JW
0131 552 3439
bronwright@btinternet.com

Dr David Wright 20 Lennox Row, Edinburgh EH5 3JW
0131 552 3439
bdwright20lr@btinternet.com

ULSTER BEEKEEPERS' ASSOCIATION

www.ubka.org

OBJECTS OF THE ASSOCIATION

The objects of the Association are to unite beekeepers for their mutual benefit to serve the best interests of beekeeping by all means within its power and to foster its healthy development.

For the purpose of achieving these objects the Association will:

- promote the formation of local Beekeepers' Associations
- disseminate information and advice about beekeeping
- provide examination facilities in the craft of beekeeping
- encourage maintenance and improvement of the beekeeping environment.

EDUCATION

In conjunction with the College of Agriculture, Food & Enterprise (CAFRE), the U.B.K.A. assists in organising classes for Preliminary, Intermediate and Senior Certificate Examinations in Beekeeping following the syllabus of the Federation of Irish Beekeepers' Associations (FIBKA).

INSURANCE

Affiliated local Associations and their individual members have access to the UBKA group public and product liability insurance scheme.

APIARY SITES

Almost all 14 local Associations and CAFRE's Greenmount Campus have access to apiary sites and, for some sites, access to observation houses provided with help from Leader 2 funding, for use in demonstrating and promoting good practice to members, schools and other interested groups.

PRESIDENT,
John Witchell
johnwitchell@gmail.com

SECRETARY,
Lorraine McBride,
29 Ballylummin Road,
Ahoghill,
BT42 2PH.
email:
ubkasecretary@gmail.com

TREASURER,
Bill Rafferty
william.rafferty@virgin.net

LECTURERS
Vanessa Drew,
40 Lacken Rd.,
Ballyroney,
Banbridge,
Down BT32 5JA

Jim Fletcher
26 Coach Road, Comber
Co.Down. BT23 5QX

Lorraine McBride
29 Ballylummin Road,
Ahoghill,
BT42 2PH
028 2587 8980.

Lecturers continued

Ethel Irvine
2 Laragh Lee
Ballycassidy
ENNISKILLEN
BT94 2JT

Lorraine McBride
29 Ballylummin Road,
Ahoghill,
BT42 2PH.

Norman Walsh
43, Edentrillick Rd
Hillsborough, Co. Down
BT26 6PG

HONEY SHOWS

Local Associations stage honey shows throughout Northern Ireland. The Northern Ireland Honey Show, hosted by the Belfast City Parks Department, is held annually in September in the Botanic Gardens Belfast.

CONFERENCE

The 73rd UBKA Annual Conference will be held on 10th – 111th March 2017 at CAFRE's Greenmount Campus, Antrim. Contact the U.B.K.A. Conference Secretary at 07871 161303 and www.ubka.org for details.

SECRETARIES OF ASSOCIATIONS

Belfast,
Jonathan Getty
80 Locksley Park,
Belfast,
BT10 0AS
email: bbkasecretary@
googlemail.com

Clogher Valley,
Chester Roulston
10 Ednagee Rd,
Garvetagh, Castlederg,
Co. Tyrone
BT81 7QF
email: chester.roulston@
hotmail.co.uk.

Derry City, Jen Simpson
4 Cullinean Manor,
Redcastle,
Nr Lifford
Co. Donegal
email: jennifersimpson163@
yahoo.com.

Dromore, Patrick Lundy
116 Dromore Road,
Ballynahinch,
Co.Down
BT24 8HK
email: patrickjlundy@
gmail.com

East Antrim, Stephen Robinson
53 Wellington Ave.,
Larne, Co Antrim
BT40 1EH
email: admin@sgrobinson.
co.uk.

Fermanagh,
Jorgen Pedderson
email: enniskillencc@
gmail.com.

Killinchy, Dawn Stocking
Ballycruttle House,
7 Tullynaskeagh Road
Downpatrick
Co. Down
BT30 7EJ
email: kbkasecretary@
gmail.com

Mid Antrim, Angela Morrow
23 Beechwood Drive
Ahogill
Ballymena
BT42 1NB
email: midantrimbka@
btinternet.com

Mid Ulster, Anne Milligan
61 Blackisland Road ,
Annaghmore,
Portadown
BT62 1NE
email: annpmilligan@
gmail.com

Randalstown, Brian Gillanders
email: hightown@live.co.uk

Roe Valley, Sandra Logan
22 Knocknougher Road,
Macosquin,
Coleraine,
Co. Londonderry.
email: alexandramlogan@
aol.com

The Three Rivers,
Daniel McMenamin
email: daniel@urney.info.

Rostrevor & Warrenpoint,
Darren Nugent,
email: biglongdarren@
hotmail.com

Honey Judges
Jim Fletcher
26 Coach Road,
Comber
BT23 5QX

Michael Young
Mileaway, Carnreagh
Road
Hillsborough
Co. Down
BT26 6LJ

Norman Walsh
43 Edentrillick Rd
Hillsborough
Co. Down
BT26 6NH

CYMDEITHAS GWENYNWYR CYMRU WELSH BEEKEEPERS' ASSOCIATION

AMCANION Y GYMDEITHAS / AIMS OF THE ASSOCIATION

- Promote and develop beekeeping in Wales
- Conduct examinations in beekeeping
- Liaise with organisations and bodies for the benefit of beekeeping in Wales

AELODAETH UNIGOL / INDIVIDUAL MEMBERSHIP

Individual membership of the WBKA is provided for persons who do not live within the areas of branch associations, and wish to support the association. Information relating to benefits and facilities provided for individual members is available from the Individual Membership Secretary.

ARHOLIADAU / EXAMINATIONS

The WBKA offer a Basic Husbandry Assessment and support those members wishing to sit the beekeeping modules and practical examinations offered through the BBKA. Information is available from the WBKA website and from the Examinations Secretary

CYNHADLEDD/ CONVENTION

At the Royal Welsh Agricultural Society's Showground, Llanelwedd. This event is normally held during Late March/ Early April. Information relating to this event is available from the Events Secretary.

YSWIRIANT / INSURANCE

All individual and fully paid up members of beekeeping associations affiliated to WBKA are covered against 'Public and Product' liability claims. All affiliated associations are covered against public liability during conventions officially organised by the association.

LLYWYDD/PRESIDENT

Tony Shaw
School House
Y Fan
LLanidloes
Powys. SY18 6NP.

CADEIRYDD/CHAIR

Jenny Shaw
Llwyn Ysgaw,
Dwyran, Llanfairpwll,
Anglesey LL61 6RH
chair@wbka.com

YMDDIRIEDOLWYR /TRUSTEES

Jenny Shaw
chair@wbka.com

Graham Wheeler
treasurer@wbka.com

Andy Ryan
Chris Clarke
Lesley Williams

John Bowles
insurance@wbka.com

YSGRIFENNYDD / SECRETARY

secretary@wbka.com

WBKA/CGC

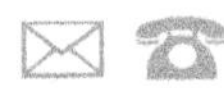

TRYSORYDD/TREASURER
Graham Wheeler

GWEFEISTR/WEBMASTER
Grant Williams

AND GOLYGYDD/EDITOR
Grant Williams
editor@wbka.com

IS-OLYGYDD (ERTHYGLAU CYMRAEG)/SUB EDITOR
Dewi Morris Jones
Llwynderw, Bronant
Aberystwyth SY23 4TG
(01974 251264)

ARHOLIADAU/EXAMINATIONS
Lynfa Davies
education@wbka.com

The WBKA Individual Membership benefits include cover under the BDI Scheme against the loss, due to foul brood diseases, of a minimum number of stocks (determined by BDI). Affiliated Associations provide this cover for their members.

GWENYNWYR CYMRU - The Welsh Beekeeper
A publication of the Welsh Beekeepers Association, giving news and views of beekeeping and related subjects. Articles and advertisements enquiries should be sent to the Editor. Gwenynwyr Cymru is provided free to members of Affiliated Associations and Individual Members. Information regarding subscriptions is available from the Individual Membership / Subscription Secretary.

CYNLLUN CYSWLLT CHWYSTRELLU / SPRAY LIAISON SCHEME
Information is available from the General secretary

SIOEAU / SHOWS
Honey/beekeeping sections are included at the Royal Welsh Agricultural Show, Llanelwedd, (OS ref: SO040520) during July, and at county, town and village shows throughout Wales. Information relating to these events may be obtained from secretaries of associations in the locality of the shows.

The historic FFAIR FEL ABERCONWY is held annually in the main street of the town, (OS ref: SH278378), on 13th September. Further information is available from the secretary of Conwy Association.

RHEOLAU CYFREITHIOL / STATUTORY REGULATIONS
The administration of the statutory regulations governing all aspects of beekeeping in Wales, is the responsibility of the Wales National Assembly, Caerdydd, CF99 1NA Phone (02920) 825111 Fax: (02920) 823352 Matters concerning statutory regulations, their implications and execution, should be addressed to the Minister of Agriculture and Rural Affairs, Wales National Assembly, at the above address.

REGIONAL BEE INSPECTOR, WALES I AROLYGYDD GWENYN
Rhanbarthol, Cymru

NATIONAL BEE UNIT I UNED GWENYN CENEDLAETHOL
ANIMAL AND PLANT HEALTH AGENCY(APHA) I ASIANTAETH IECHYD ANIFEILIAID A PHLANHIGION
07775 119480
francis.gellatly@apha.gsi.gov.uk

NATIONAL BEE UNIT WEBSITE (BEEBASE)
www.nationalbeeunit.com
National Bee Unit, National Agri-Food Innovation Campus, Sand Hutton, York, YO41 1LZ

OFFICE OF THE CHIEF VETERINARY OFFICER / SWYDDFA'R PRIF SWYDDOG MILFEDDYGOL
Agriculture, Veterinary, Food and Marine – Amaeth, Milfeddygaeth, Bwyd a'r Môr

AELODAETH UNIGOL-TANYSGRIFAU/ MEMBERSHIP - INDIVIDUAL MEMBERSHIP SUBSCRIPTIONS
Ian Hubbuck
White Cottage
Berriew
SY21 8BB
01686 640205
ianhubbuck@hotmail.com

INSURANCE:
John Bowles
insurance@wbka.com

EVENTS/CONVENTION SECRETARY:
Jill Wheeler
mertyndowning@
btinternet.com

WBKA/CGC

CYMDEITHASAU TADOGOL A'U YSGRIFENYDDION / AFFILIATED ASSOCIATIONS AND SECRETARIES

ABERYSTWYTH, Ann Ovens,
Tan-y-Cae, Nr Talybont,
Ceredigion,
SY24 5DP
01970 832359
ann.ovens@btinternet.com

ANGLESEY, John Bowles,
Penrhos
Llanfaglan
Caernarfon
LL54 5RB
07449 021219
secretaryabka@gmail.com

BRECKNOCK AND RADNOR,
Roger James
Ty Siloh,
Llandeilo'R Fan,
BRECON
LD3 8UD
gwenynwyrbrycheiniog@
gmail.com

BRIDGEND, Shirley Myall
Distant View
Penylan Road
St Brides Mayor
Bridgend
CF32 0SA
shirley.myall@btinternet.com

CARDIFF AND VALE,
Annie Newsam
Stonecroft, Mountain Road,
Bedwas, Caerphilly, CF83 8ER
annienewsam@hotmail.co.uk

CARMARTHEN, Stephen Cox
Pen-Y-Maes
Ostrey Hill
St Clears
SA33 4AJ
steve_p_cox30@hotmail.com
07906 515996

CONWY, Mr Peter McFadden,
Ynys Goch
Ty'n y Groes,
Conwy LL32 8UH
01492 650851
peter@honeyfair.freeserve.co.uk

EAST CARMARTHEN
Doug Taylor,
Rhydgoch
Golden Grove
Carmarthen
SA32 8ND
01558 668339
doug@rhydgoch.co.uk

FLINT AND DISTRICT,
Jill and Graham Wheeler,
Mertyn Downing, Whitford
Holywell, Flintshire,
CH8 9EP.
01745 560557
mertyndowning@btinternet.
com

GWENYNWYR CYMRAEG
CEREDIGION W.I.Griffiths,
Llain Deg, Comins Coch,
Aberystwyth, SY23 3BG
01970 623334
wilmair@btinternet.com

LAMPETER AND DISTRICT
Jan Crabb,
Llwyn Myfryd,
Llandewi Brefi,
TREGARON
SY25 6SB
jancrabb@hotmail.com

LLEYN AC EIFIONYDD
Amanda Bristow,
Bryngwydion, Pontllyfni,
Gwynedd
LL54 5EY 01286 831328
amanda.bristow@egnitec.com

MEIRIONNYDD,
Sue Davies
Nant Ddu
Arenig
Bala
LL23 7PA
suedavies0117@gmail.com
01766 540790

MONTGOMERYSHIRE,
Helen Woodruff
Manledd,
LLANIDLOES
SY18 6NP
secretary@montybees.org.uk

PEMBROKESHIRE,
Lesley Williams
Tanffynnon
Blaenffos
Boncath
SA37 0HT

SOUTH CLWYD,
David P. Evans
Cwellyn
Dinmael
Corwen
LL21 0NY
dinmaelbees@btinternet.com
01490 460329

SWANSEA,
Julian Caruana
61 Glanyrafon Road
Pontarddulais
Swansea
SA4 8LT
07985 328910
sdbks.secretary@gmail.com

TEIFISIDE
Marion Dunn
Wenallt,
Beulah,
NEWCASTLE EMLYN
SA38 9QH
secretary@tbka.org.uk

WEST GLAMORGAN
John Beynon
48 Whitestone Avenue
Bishopston
Swansea
SA3 3DA
jakbeynon@btinternet.com

BEIRNIAID SIOE FÊL TRWYDDEDIG / WBKA QUALIFIED HONEY SHOW

TERRY E. ASHLEY
Meadow Cottage,
11 Elton Lane, Winterley
Sandbach CW11 4TN
M. J. BADGER MBE
14 Thorn Lane, Leeds
LS8 1NN
M BESSANT
Gwili Lodge, Heol
Lotwen, Rhydaman
SA18 3RP
ROBERT BREWER
PO Box 369, Hiawassee,
Georgia, USA
TOM CANNING
151 Portadown Road,
Armagh, Co Armagh
BT61 9HL
LES CHIRNSIDE
Bryn-y-Pant Cottage,
Upper Llanover,
Abergavenny NP7 9ES
CARYS EDWARDS
Ty Cerrig, Ganllwyd,
Dolgellau LL40 2TN
IFOR C. EDWARDS
Lleifior, Pontrhydygroes,
Ystrad Meurig SY25 6DN
STEVEN GUEST
Bridge House, Hind
Heath Road, Sandbach,
CW11 3LY
HUGH MCBRIDE
11 Ballyloughan Park
Antrim BT43 5HW
LORRAINE MCBRIDE
11 Ballyloughan Park
Antrim BT43 5HW
CECIL MCMULLAN
33 Glebe Road,
Hillsborough, County
Down
LEO MCGUINESS
89 Dunlade Road, Grey
Steel BT47 4QL
GAIL ORR
64 Ballycrone Road,
Hillsborough BT26 6NH
DINAH SWEET
Graig Fawr Lodge,
Caerphilly, CF83 1NF
REDMOND WILLIAMS
Tincurry, Cahir, Co
Tipperary Eire
MICHAEL YOUNG MBE
Mileaway, Carnreagh,
Hillsborough BT26 6LJ

NATIONAL BEE UNIT, Animal and Plant Health Agency (APHA)

www.nationalbeeunit.com

National Bee Unit
APHA, National Agri-Food Innovation Campus
Sand Hutton, York,
YO41 1LZ, UK

Tel.No: 0300 3030094
Fax.No: 01904 462240
E-Mail: nbu@apha.gsi.gov.uk
Website: www.nationalbeeunit.com

NATIONAL BEE UNIT TECHNICAL STAFF, HEAD OF UNIT
Mike Brown

NATIONAL BEE INSPECTOR
Andy Wattam
01522 789726
07775 027524

CONTINGENCY PLANNING
Nigel Semmence
01264 338694
07776 493649

National Bee Unit

The National Bee Unit (NBU) is part of the Animal and Plant Health Agency (APHA), an agency of the Department for Environment, Food and Rural Affairs (Defra), and is based just outside of York. The NBU is an element of APHA and its work covers all aspects of bee health and husbandry in England and Wales, on behalf of Defra in England and for the Welsh Government in Wales. The work of the unit includes disease and pest diagnosis, research into bee health matters, development of contingency plans for emerging threats, import risk analysis, related extension work and consultancy services to both Government and industry.

Inspection work is accredited by UKAS to ISO 17020.

Bee Health Inspection Service

The NBU has a long track record in bee husbandry and bee disease control (since 1946) and has been directly responsible for the Bee Inspection services in England and Wales since 1994. The NBU consists of a home-based Inspectorate team, and the laboratory diagnostic and research team based at National Agri-Food Innovation Campus, Sand Hutton. In addition colleagues across Fera contribute to the programme and research projects. The Bee Health Inspectorate team consists of approximately 60 home-based members of staff. It is headed by the National Bee Inspector (NBI), whose role it is to manage the statutory disease control and training programmes. The NBI has management responsibility for eight home-based Regional Bee Inspectors (RBIs), one heading each of the seven regions in England and one covering Wales. The RBI in turn manages a

number of Seasonal Bee Inspectors (SBIs). The RBIs and SBIs organise inspections under EU and UK legislation, submit suspect samples for diagnosis, treat colonies for foulbrood and train beekeepers in bee husbandry for better disease control and greater self-sufficiency. In addition the Bee Inspectors also collect honey samples for residue analysis under the Statutory Honey collection agreement with Defra Veterinary Medicines Directorate (VMD). With *Aethina tumida* (Small hive beetle (SHB)) and *Tropilaelaps spp.* both notifiable under UK and EU law, Inspectors also undertake surveillance for these exotics in "Sentinel Apiaries (SA)" close to identified high risk areas. Beekeepers who manage SA's represent a valuable front line defence against an exotic pest incursion.

Bee Disease Diagnostic Team

The NBU's diagnostic team provides a rapid, modern service for both the Inspection team and beekeepers. The NBU laboratory adheres to Good Laboratory Practice (GLP) to ensure a high professional standard. All diagnostic tests are conducted according to the OIE (Office International des Epizooties) Manual of Standard Diagnostic Tests and Vaccines. The OIE is the World Organisation for Animal Health and produce internationally recognised disease diagnosis guidelines (http//www.oie.int.). Across Fera, diagnostic support is provided from teams of microbiologists, acarologists, insect virologists and molecular specialists.

Bees and the Law

The 1980 Bees Act empowers Ministers to make Orders to control pests and diseases affecting bees, and provides powers of entry for Authorised Persons. Under the Bees Act, The Bee Diseases and Pests Control Order 2006 for England (as amended) and Wales, (there is similar legislation for Scotland and Northern Ireland) designates American foulbrood (AFB), European foulbrood (EFB), *A. tumida* (SHB) and *Tropilaelaps* mites (all species) as notifiable pests and defines the action which may be taken in the event of outbreaks. The Trade in Animals and Related Products Regulations 2011 (TARP) give effect to EU legislation

REGIONAL BEE INSPECTORS

Ian Molyneux
Northern Region
01204 381186
07775 119442

Jo Schup
Western Region
01948 710731
07979 119368

Julian Parker
Southern Region
01494 488393
07775 119469

Diane Steele
South East Region
01243 582612
07775 119452

Simon Jones
South West Region
01823 442228
07775 119459

Keith Morgan
Eastern Region
01485 520838
07919 004215

Ivor Flatman
North East Region
01924 252795
07775 119436

Frank Gellatly
Wales
01558 650663
07775 119480

To find details of Seasonal / Bee Inspectors please see BeeBase: https://secure.fera.defra.gov.uk/beebase/index.cfm . Remember that Seasonal Inspectors only work from April to September

Research Scientist
Kirsty Stainton

Laboratory
Ben Jones (laboratory manager)
Thomas Broadhead
Victoria Tomkies

Apiarists
Jack Wilford

Technical Advisor
Jason Learner

Administrative Programme Support
Kate Parker,
Lesley Debenham
& Jenna Cook

concerning trade in animals and animal products from other Member States and the importation of animals and animal products from Third Countries. Consolidated in TARP is the Directive on animal health requirements for trade in bees (the Balai Directive (92/65/EEC)). The Balai Directive specifically describes the intra-trade certification requirements for honey bees. Annexes A and B list the pests and diseases of animals, including those that affect honey bees, which are considered highest risk. Annex A lists the notifiable organisms throughout the Union. These are American foulbrood (AFB), the Small hive beetle (*A. tumida*) and *Tropilaelaps* mites. Annex B lists organisms that are not notifiable across the EU, but which Member States may choose to cover under their own domestic legislation. At the time of writing *Tropilaelaps* has not been confirmed in Europe.

The Importation of Bees

Beekeepers may legally import Queen honey bees from listed Third Countries as set out in the Commission Regulation (EU) 206/2010 . There are three countries outside of the EU, that we are aware of, whom can comply with the Decision: Argentina, Australia and New Zealand. Honey bee queens and packages may be imported from the EU, however, the beekeeper must make themselves aware of and adhere to the importation guidelines which can be found on BeeBase: http://www.nationalbeeunit.com/index.cfm?sectionid=47.

Under the Balai directive consignments of bees moved between Member States must be accompanied by an original health certificate confirming freedom from notifiable pests and diseases. In addition, colonies may be subject to controls aimed at preventing the spread of Fireblight between 15th March and 30th June. More information about this can be found on BeeBase http://www.nationalbeeunit.com/index.cfm?pageId=103

American and European foulbrood

American and European foulbrood are both serious diseases of European honey bees and are subject to statutory control in the United Kingdom.

Any beekeeper who suspects their colonies to be infected with foulbrood should contact their nearest Appointed Bee Inspector (ABI) to report this.

Those apiaries that are suspected or confirmed to have a notifiable disease will be issued with a Standstill notice prohibiting the movement of any hive, bees, combs, bee products, bee pests, hive debris, appliances or other things liable to spread a suspected notifiable disease or pest on those premises or vehicle except under licence. More information about foulbrood disease is available on our website or in our advisory leaflet 'Foulbrood Disease of Honey Bees and other common brood disorders' http://www.nationalbeeunit.com/index.cfm?pageid=167.

Varroa

As part of the NBU's routine field screening programme the first known case of pyrethroid resistant Varroa mites in the UK was discovered in apiaries in Devon in August 2001. The NBU undertook a resistance-monitoring programme throughout England and Wales. Pyrethroid resistant *Varroa* mites are now widespread in England and Wales. To access current advice on *Varroa* and it's management please visit BeeBase.

Exotics
Beekeepers must make themselves aware of the potential threats to beekeeping in the UK. The field Inspection team monitors for potential exotics, the SHB and *Tropilaelaps spp*. The laboratory team also routinely screens import and suspect samples submitted for identification by both beekeepers and the field team.

Pesticide Monitoring
The WIIS involves the collaborative work of four separate organisations. It is led by the Chemicals Regulation Directorate (CRD) "http://www.pesticides.gov.uk/", formerly the Pesticides Safety Directorate. They are the 'Competent Authority' for the approval and regulation of pesticides and some other chemicals. Natural England (NE) "https://www.gov.uk/government/organisations/natural-england" manage the Scheme on the behalf of CRD and undertake site enquiries into pesticide exposure and, Fera Science Ltd (Fera) carry out pesticide analysis and, if appropriate, theAnimal and Plant Health Agency (APHA) "https://www.gov.uk/government/organisations/animal-and-plant-health-agency" carry out post mortems on wildlife. At Fera the Wildlife Incident Unit (WIU) "http://fera.co.uk/ccss/WIIS.cfm" will analyse samples for pesticides. and give an interpretation of the result based on information from the agencies involved.

It can also result in changes to label recommendations on pesticide products. It is not provided as a personal service to beekeepers wishing to seek evidence for the purpose of civil litigation but can lead to enforcement action being taken by the enforcer if the misuse or abuse of a product is identified as part of this process. For more information please see BeeBase: https://secure.fera.defra.gov.uk/beebase/index.cfm?pageId=84.

Research & Development
A programme of research and development within the group underpins the Unit's work. They also have long-established links with many European and world wide research centres, universities and the beekeeping industry. The primary aim of our R&D is to improve our understanding of the issues which impact bee health. The NBU also actively supports PhD students, some of which are funded using donations from the beekeeping industry.
For an update on the current R&D work of the unit please see BeeBase.

Extension

The NBU trains beekeepers in several ways: local courses and advisory visits run by the Inspectors, and national courses held at the York laboratory. The NBU annually hosts the National Diploma in Beekeeping residential courses and has also been host to visiting overseas workers and researchers. NBU York based staff also provide training to beekeepers at local and regional beekeeper meetings.

Healthy Bees Plan

The Healthy Bees Plan was published by Defra and the Welsh Assembly Government in March 2009 following consultation with beekeepers and the main Beekeeping Associations. It sets out a plan for Government, beekeepers and other stakeholders to work together to respond effectively to pest and disease threats and to put in place programmes to ensure a sustainable and productive future for beekeeping In England and Wales. The Healthy Bees Plan consists of three working groups that report to the project management board to help deliver the five major objectives of the plan. To view the Healthy Bees Plan, please see BeeBase.

BeeBase

BeeBase is the National Bee Unit website. It is designed for beekeepers and supports Defra, WAG and Scotland's Bee Health Programmes and the Healthy Bees Plan, which set out to protect and sustain our valuable national bee stocks. Our website provides a wide range of free information for beekeepers, to help keep their honey bees healthy. We hope both new and experienced beekeepers will find this an extremely useful resource and sign up to BeeBase. Knowing the distribution of beekeepers and their apiaries across the country helps us to effectively monitor and control the spread of serious honey bee pests and diseases, as well as provide up-to-date information in keeping bees healthy and productive. By telling us who you are you'll be playing a very important part in helping to maintain and sustain honey bees for the future. To register as a beekeeper please visit www.nationalbeeunit.com.

BEE HEALTH INSPECTIONS:
Thomas Williamson
Plant Health Inspection Branch,
DAERA,
Atek Building, Edenaveys
Industrial Estate,
Newry Road,
Armagh,
BT60 1NF
Tel: 028 3889 2374
Email: tom.williamson@
daera-ni.gov.uk

BEE DISEASE DIAGNOSTICS:
Ivan Forsythe
Agri-Food and Biosciences
Institute (AFBI)
Newforge Lane
BELFAST BT9 5PX
Tel: 028 9025 5288
Email:
Ivan.Forsythe@afbini.gov.uk

Honeybee Regional Report for Northern Ireland 2016

Bee Health Inspections

Following the increased levels of American foulbrood (AFB) and European foulbrood (EFB) found last year in Northern Ireland, DAERA Bee Health Inspectorate have continued their inspection and surveillance programme in support of local beekeepers. To date 90 beekeepers have received an inspection which has resulted in confirmation of AFB at 13 apiaries and EFB at 3 apiaries. A regularly updated disease findings map has been published on DAERA web pages detailing the approximate locations of outbreaks, in order to assist beekeepers to be more aware of any increased localised risk of disease to their colonies. Surveys for Small hive beetle and Tropilaelaps mites have also continued to be part of the inspection programme in conjunction with the sentinel apiary programme. Recognising the need for additional suitably qualified seasonal Bee Inspectors to support the bee health inspection functions, DAERA are training new inspectors to address the increase in brood disease findings in recent seasons.

Bee Health Surveys

A questionnaire survey for Bee Husbandry issues has been circulated annually to beekeepers via beekeeping associations since 2009. The results of the surveys up until 2013 are available as a pdf file on the AFBI website (www.afbini.gov.uk). Preliminary results from the 2016 survey show that beekeepers recorded 31% of their overwintering colonies lost due to death/queen problems. 41% of responding beekeepers reported no losses. The 2015 survey results are currently being processed but soon will be available on the AFBI website.

Adult Bee Disease Diagnostics

Up to September 2016, 74 samples of bees had been examined for Nosema infections. 36 (49%) were positive for Nosema spp. These results are from microscopic examination. Similarly in 2013, 49% of samples were found to be positive for Nosema. Note, this should not be used as an indicator of prevalence, as samples were not spatially representative of colony distribution in Northern Ireland.

Sentinel Apiaries

Inspectors, with the support of local beekeepers, have established sentinel apiaries in support of the early detection of the quarantine pests Small hive beetle and Tropilaelaps mites. These have been located at identified high risk areas throughout Northern Ireland at Association and beekeepers apiaries.

Disease Identification Workshops

DAERA has been hosting practical bee disease identification workshops to help beekeepers develop their skills in identifying brood diseases. Organised in conjunction with the Ulster Bee Keepers Association (UBKA) and Institute of Northern Ireland Beekeepers (INIB) the workshops were attended by beekeepers, Association Bee Health Officers, Lecturers, Apiary Managers and Mentors. Beekeepers attending these events have commented very favourably on the benefits of these and they have already resulted in cases of notifiable diseases being reported.

Thomas Williamson
Plant Health Inspection Branch, DAERA
Ivan Forsythe
Bee Disease Diagnostics, AFBI

THE SCOTTISH GOVERNMENT AGRICULTURE, FOOD AND RURAL COMMUNITIES DIRECTORATE (AFRC) - RURAL PAYMENTS AND INSPECTIONS DIRECTORATE (RPID)

HEADQUARTERS
Lead Bee Inspector
Stephen Sunderland,
P Spur, Saughton House,
Broomhouse Drive,
Edinburgh, EH11 3XD
Tel: 0300 244 6672
e-mail:
bees_mailbox@gov.scot

The Scottish Government (SG) is responsible for bee health Policy in Scotland. SG recognises the importance of a strong Bee health programme, not only for the production of honey, but also for the contribution that bees make to the pollination of many crop species and to the wider environment.
Honey bees are susceptible to a variety of threats, including pests and diseases, the likelihood and consequences of which have increased significantly over the last few years.

The Scottish Government takes very seriously any biosecurity threat to the sustainability of the apiculture sector and is working closely with colleagues in Food and Environment Research Agency's (Fera) National Bee Unit (NBU) to enable a more joined up approach to be taken throughout Great Britain on the issues surrounding bee health.

The Scottish Government has invested in the NBU's National web-based database for beekeepers "BeeBase" and actively encourages beekeepers to register onto the system. This service will provide bee health and disease outbreak information and will also assist Bee Inspectors in disease control. BeeBase also provides information on legislation, pests and disease recognition and control, interactive maps, current research areas and key contacts.

Beekeepers have a significant role to play in ensuring disease management and control within their own apiaries are in order as they have a legal obligation to report any suspicion of a notifiable disease or pest to the Bee Inspector at their local SGRPID Area Office. Bee Inspectors are responsible for the operation of The Bee Diseases and Pests Control (Scotland) Order 2007 in their area with duties including:-

- Inspection of apiaries for presence of statutory bee diseases
- Taking and delivering samples to SASA
- Issuing and removal of 'Standstill Notices'
- Issuing of 'Destruction Notices' and supervising destruction
- Informing beekeepers of treatment options for European Foul Brood (EFB), where appropriate
- Granting the option, after taking account of the recommendations of SASA, and carrying out treatment
- Carrying out follow-up inspections after destruction or treatment

SASA

- **Science and Advice for Scottish Agriculture (SASA)** is responsible for providing specialist technical support where duties include:
- Examination of submitted samples suspected of being infected with American Foul Brood, European Foul Brood, Small Hive Beetle (SHB) or *Tropilaelaps*.
- Reporting results on which pathogen or pest is present
- Recommending, in consultation with the Bee Inspector, the most suitable option, destruction or treatment, for each individual case of EFB.
- Where treatment is agreed, ordering supplies of the approved antibiotic
- Provision of a free diagnostic service to beekeepers to identify and confirm the presence of varroa.
- Maintaining technical liaison with NBU and providing technical documentation as required
- Providing training courses and demonstration material as required

SASA (SCIENCE AND ADVICE FOR SCOTTISH AGRICULTURE)
1 Roddinglaw Rd,
Edinburgh, EH12 9FJ

BEE DISEASES, FIONA HIGHET
Virology and Zoology Section
(0131) 244 8817

PESTICIDE INCIDENTS, ELIZABETH SHARP
Pesticides Section
(0131) 244 8874

PESTICIDE INCIDENTS

As part of the Wildlife Incident Investigation Scheme (WIIS), SASA undertakes analytical investigations into bee mortalities where pesticide poisoning may have been involved. Beekeepers should send samples of dead bees (200) direct to SASA, Pesticides Section, for analysis. In the case of major incidents, beekeepers are advised to contact their nearest SGRPID Area Office so that an early field investigation can be instigated.

THE FOLLOWING SCOTTISH GOVERNMENT RURAL PAYMENTS AND INSPECTIONS DIRECTORATE (SGRPID) STAFF ARE AUTHORISED BEE INSPECTORS. ALL BEE INSPECTORS HAVE EMAIL ADDRESSES AS "FIRSTNAME.SURNAME@SCOTLAND.GOV.SCOT"

EDINBURGH (HQ)
Steve Sunderland
(Lead Bee Inspector)
P Spur, Saughton House,
Broomhouse Drive,
Edinburgh, EH11 3XD
Tel: 0300 244 6672
Fax: 0300 244 9797
steve.sunderland@gov.scot

GRAMPIAN (INVERURIE AREA OFFICE)
Kirsteen Sutherland
Thainstone Court,
Inverurie, Grampian,
Aberdeenshire, AB51 5YA
Tel: (01467) 626247
Fax: (01467) 626217
kirsteen.sunderland@gov.scot

HIGHLAND (INVERNESS AREA OFFICE)
Gordon Mackay
Longman House
28 Longman Road
Inverness, IV1 1SF
Tel: 01463 253 053
gordon.mackay@gov.scot

SOUTH EASTERN (GALASHIELS AREA OFFICE)
Angus MacAskill
Cotgreen Road
Tweedbank, Galashiels
Scottish Borders, TD1 3SG
Tel: (01896) 892400
Fax: (01896) 892424
angus.macaskill@gov.scot

SOUTH WESTERN (AYR AREA OFFICE)
John Smith
Russell House
King Street
Ayr
South Ayrshire
KA8 0BG
Tel: (01292) 291300
Fax: (01292) 291301
john.smith@gov.scot

SCOTLAND'S RURAL COLLEGE (SRUC)

The Scottish Government supports a full-time apiculture specialist (Graeme Sharpe) who provides comprehensive advisory, training and education programmes for beekeepers throughout Scotland on all aspects of Integrated Pest Management, good husbandry (including the control of Varroa) and management practices. SAC also promotes the awareness of notifiable bee diseases and pests.

GRAEME SHARPE, APICULTURE SPECIALIST,
SAC Consulting, Scotland's Rural College (SRUC),
Veterinary Services
John Niven Building
Auchincruive Estate
Ayr, Ayrshire
KA6 5HW
Tel:01292 525375

WWW.SCOTLAND.GOV.UK/TOPICS/APICULTURE/GRANTS/INSPECTIONS/BEEINSPECTIONS

USEFUL TABLES

BEEKEEPING METRIC CONVERTION TABLES

°CENT	FAHR	INCH	MM	INCH	MM	INCH	MM
0	32	1/25	1	1 5/8	42	10	254
5	40	1/12	2	1 11/16	43	10 1/4	260
7	44	1/8	3	1 9/20	48	11 1/4	286
30	86	1/16	5	2	51	11 1/2	292
34	92	1/4	6	3	76	11 3/4	298
38	100	5/16	8	4 1/4	108	12	305
43	110	3/8	9	4 1/2	114	14	356
49	120	1/2	12.5	4 3/4	121	16 1/4	413
54	130	5/8	16	5 1/2	140	16 1/2	49
60	140	3/4	18	5 3/4	146	17	431
62	144	7/8	22	6	152	17 5/8	448
82	180	1	25	6 1/4	159	18 1/8	460
90	194	1 1/16	27	8 1/4	216	18 1/4	483
100	212	1 3/8	35	8 3/4	223	20	508
		1 9/20	37	9 1/8	232	21 1/2	546
		1 1/2	38	9 3/8	239	21 3/4	552
				9 9/16	246	22	559

INTERNATIONAL QUEEN MARKING COLOURS

YEAR ENDING	COLOUR	REMEMBER
1 & 6	WHITE	Will
2 & 7	YELLOW	You
3 & 8	RED	Raise
4 & 9	GREEN	Good
5 & 0	BLUE	Bees?

USEFUL TABLES

BOTTOM BEE-SPACE HIVES

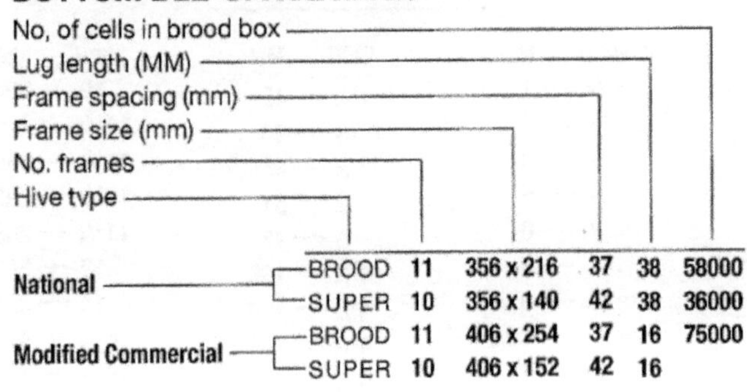

Hive tvpe		No. frames	Frame size (mm)	Frame spacing (mm)	Lug length (MM)	No, of cells in brood box
National	BROOD	11	356 x 216	37	38	58000
	SUPER	10	356 x 140	42	38	36000
Modified Commercial	BROOD	11	406 x 254	37	16	75000
	SUPER	10	406 x 152	42	16	

TOP BEE-SPACE HIVES

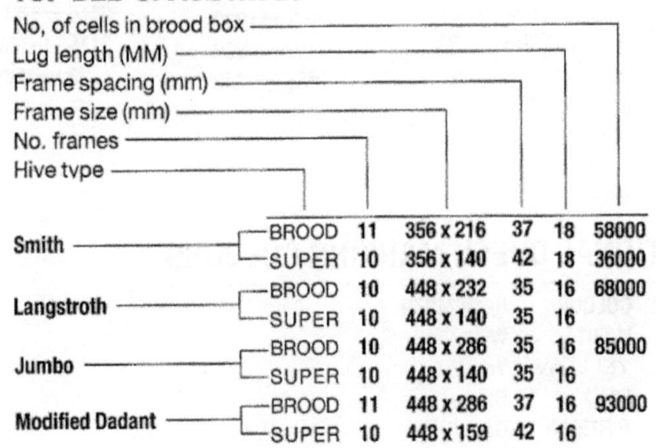

Hive tvpe		No. frames	Frame size (mm)	Frame spacing (mm)	Lug length (MM)	No, of cells in brood box
Smith	BROOD	11	356 x 216	37	18	58000
	SUPER	10	356 x 140	42	18	36000
Langstroth	BROOD	10	448 x 232	35	16	68000
	SUPER	10	448 x 140	35	16	
Jumbo	BROOD	10	448 x 286	35	16	85000
	SUPER	10	448 x 140	35	16	
Modified Dadant	BROOD	11	448 x 286	37	16	93000
	SUPER	10	448 x 159	42	16	

USEFUL TABLES

CONVERSION FACTORS

TEMPERATURE

Fahrenheit > Celcius (Centigrade)	- 32, x 0.5555 (5/9)
Celcius > Fahrenheit	x 1.8 (9/5), + 32

WEIGHT

Ounces > Pounds	x 28.3495
Pounds > Grams	x 453.59237
Hundredweights > Kilograms	x 50.8
Grams > Ounces	÷ 28.3495
Kilograms > Pounds	x 2.2142

LENGTH

Inches > Centimetres	x 2.54
Yards > Metres	x 0.9144
Miles > Kilometres	x 1.609
Centimetres > Inches	x 0.3937
Metres > Yards	x 1.0936
Kilometres > Miles	÷ 1.609

AREA

Acres > Hectares	x 0.404686
Hectares > Acres	x 2.47105

VOLUME

Pints > Litres	x 0.5683
Gallons > Litres	x 4.546
Litres > Pints	x 1.7598
Litres > Gallons	x 0.21997

WORD SEARCH ANSWERS

ABDOMEN
ANTENNA
AURICLE
CIBARIUM
CORBICULA
CROP
ENDOPHALLUS
GALEAE
HAEMOLYMPH
HAMULI
HYPOPHARYNGIAL
JOHNSTON
KASCHEVNIKOV
LABELLUM
LABRUM
MALPIGHIANTUBULES
MANDIBLES
NASANOV
OCELLI
PEDICEL
PROBOSCIS
PROVENTRICULUS
RASTELLUM
SENSILLA
SETAE
SPERMATHECA
SPIRACLES
THORAX
TRACHEAE
VENOMSAC

Lightning Source UK Ltd.
Milton Keynes UK
UKHW02f0633111018
330371UK00014B/1083/P

9 781908 904928